T0309385

Diversities in Quantum Computation and Quantum Information

Kinki University Series on Quantum Computing

Editor-in-Chief: Mikio Nakahara *(Kinki University, Japan)*

ISSN: 1793-7299

Published

Vol. 1   Mathematical Aspects of Quantum Computing 2007
edited by Mikio Nakahara, Robabeh Rahimi (Kinki Univ., Japan) &
Akira SaiToh (Osaka Univ., Japan)

Vol. 2   Molecular Realizations of Quantum Computing 2007
edited by Mikio Nakahara, Yukihiro Ota, Robabeh Rahimi, Yasushi Kondo &
Masahito Tada-Umezaki (Kinki Univ., Japan)

Vol. 3   Decoherence Suppression in Quantum Systems 2008
edited by Mikio Nakahara, Robabeh Rahimi & Akira SaiToh
(Kinki Univ., Japan)

Vol. 4   Frontiers in Quantum Information Research: Decoherence, Entanglement,
Entropy, MPS and DMRG 2009
edited by Mikio Nakahara (Kinki Univ., Japan) &
Shu Tanaka (Univ. of Tokyo, Japan)

Vol. 5   Diversities in Quantum Computation and Quantum Information
edited by edited by Mikio Nakahara, Yidun Wan & Yoshitaka Sasaki
(Kinki Univ., Japan)

Vol. 6   Quantum Information and Quantum Computing
edited by Mikio Nakahara & Yoshitaka Sasaki (Kinki Univ., Japan)

Vol. 7   Interface Between Quantum Information and Statistical Physics
edited by Mikio Nakahara (Kinki Univ., Japan) &
Shu Tanaka (Univ. of Tokyo, Japan)

Vol. 8   Lectures on Quantum Computing, Thermodynamics and Statistical Physics
edited by Mikio Nakahara (Kinki Univ., Japan) &
Shu Tanaka (Univ. of Tokyo, Japan)

Kinki University Series on Quantum Computing – Vol. 5

editors

Mikio Nakahara
Kinki University, Japan

Yidun Wan
Kinki University, Japan

Yoshitaka Sasaki
Kinki University, Japan

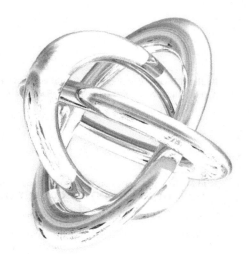

Diversities in Quantum Computation and Quantum Information

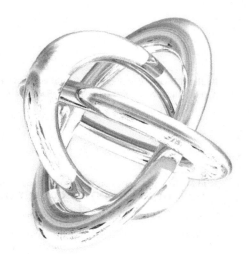 **World Scientific**

NEW JERSEY · LONDON · SINGAPORE · BEIJING · SHANGHAI · HONG KONG · TAIPEI · CHENNAI

Published by

World Scientific Publishing Co. Pte. Ltd.
5 Toh Tuck Link, Singapore 596224
USA office: 27 Warren Street, Suite 401-402, Hackensack, NJ 07601
UK office: 57 Shelton Street, Covent Garden, London WC2H 9HE

British Library Cataloguing-in-Publication Data
A catalogue record for this book is available from the British Library.

Kinki University Series on Quantum Computing — Vol. 5
DIVERSITIES IN QUANTUM COMPUTATION AND QUANTUM INFORMATION

Copyright © 2013 by World Scientific Publishing Co. Pte. Ltd.

All rights reserved. This book, or parts thereof, may not be reproduced in any form or by any means, electronic or mechanical, including photocopying, recording or any information storage and retrieval system now known or to be invented, without written permission from the Publisher.

For photocopying of material in this volume, please pay a copying fee through the Copyright Clearance Center, Inc., 222 Rosewood Drive, Danvers, MA 01923, USA. In this case permission to photocopy is not required from the publisher.

ISBN 978-981-4425-97-1

Printed in Singapore.

PREFACE

This volume contains lecture notes and poster contributions presented at the Summer School on Diversities in Quantum Computation/Information, held in Osaka from August 1 to August 5, 2010. It was intended to bring students and researchers to recent interdisciplinary developments in Quantum Computation nand Information.

The summer school started with the lecture by Prof. Li, who gave an introduction to the connections between quantum operations/channels and completely positive linear maps. The next lecture was given by Prof.Marx. He described experimental results for a small-scale approximate evaluation of the Jones polynomials by NMR. The afternoon session in the day started with the lecture by Prof. Fuentes who described relativistic effects on entanglement in flat and curved spacetimes. The lecture of Prof. van Dam started from the third day. He presented the current status of quantum algorithms with an algebraic flavor. The afternoon session in the fourth day and the last day were mainly devoted for tutorial of the lectures. Various exercises presented in the lectures were expounded mainly by graduate students and postdoctoral fellows. Moreover, there were eight contributed talks on topological quantum computation, entanglement and so on. There were active discussions during the summer school. It surely bred a congenial atmosphere to learn various topics presented in the summer school.

This summer school was supported financially by "Open Research Center" Project for Private Universities: matching fund subsidy from MEXT (Ministry of Education, Culture, Sports, Science and Technology).

We would like to thank all the lecturers and participants, who made this summer school invaluable. We would like to thank Ryota Mizuno for TeXnical assistance. Finally, we would like to thank Zhang Ji and Rhaimie Wahap of World Scientific for their excellent editorial work.

Mikio Nakahara
Yidun Wan
Yoshitaka Sasaki
Osaka, January 2012

Summer School on
Diversities in Quantum Computation/Information

Kinki Univerisity, Osaka, Japan

---

1 – 5   August   2010

---

1 August

Chi-Kwong Li (College of William and Mary, USA)
Operator approach to quantum operations, quantum channels, and quantum error correction 1

Raimund Marx (Technische Universität München, Germany)
Untying Knots by NMR 1

Ivette Fuentes (University of Nottingham, UK)
Relativistic Quantum Information 1

Shu Tanaka (Kinki University, Japan)
Quantum annealing from a viewpoint of effect of some kind of fluctuation for Potts model

Masamitsu Bando, (Kinki University, Japan)
Geometric Quantum Gates, Composite Pulses, and Trotter-Suzuki Formulas

Ayumu Sugita, (Osaka City University, Japan)
Multipartite entanglement measures in terms of group representation theory

Chi-Kwong Li (College of William and Mary, USA)
Operator approach to quantum operations, quantum channels, and quantum error correction 2

2 August

Raimund Marx (Technische Universität München, Germany)
Untying Knots by NMR 2

Chi-Kwong Li (College of William and Mary, USA)

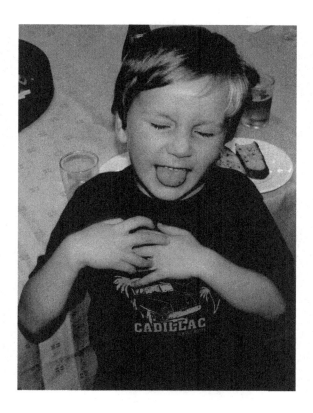

LIST OF PARTICIPANTS

Assad Hafiz	COMSATS Institute of Information Technology Park Road
Akira SaiToh	Kinki University
Ayumu Sugita	Osaka City University
Chi-Kwong Li	College of William and Mary
Chiara Bagnasco	Kinki Daigaku
Elham Hosseini	Kinki University
Eyuri Wakakuwa	the University of Tokyo
Hideo Miyata	Kanazawa Institute of Technology
Hiroyuki Miyoshi	Kyoto Sangyo University
Hiroyuki Tomita	Kinki University
Hossein Mehri Dehnavi	Kinki University
Ivette Fuentes	University of Nottingham
Karim Elgammal	Nile University
Masamitsu Bando	Kinki University
Masayuki Imai	Interdisciplinary school of scinence and engineering, Kinki University
Meiro Chiba	Kinki University
Mikio Nakahara	Kinki University
Minoru Koga	Tokyo Institute of Technology
Mohammad Ali Fasihi	Kinki University
Mustafa Sahin	Canakkale Onsekiz Mart Universty
Raimund Marx	Technische Universität München
Satoru Ishijima	Kinki University
Shogo Tanimura	Kyoto University
Shu Tanaka	Kinki University
Toshihiko Sasaki	The University of Tokyo
Tsubasa Ichikawa	Kinki University
Wim van Dam	University of California at Santa Barbara
Yasuo Ohno	Kinki University
Yasushi Kondo	Kinki University
Yidun Wan	Kinki University
Yoshitaka Sasaki	Kinki University
Yuji Tanaka	Osaka University
Yukihiro Ota	Japan Atomic Energy Agency
Yusuke Hayashi	Kyushu University
Jason Doukas	Yukawa Institute for Theoretical Physics
Ling-Yan Hung	Perimeter Institute

CONTENTS

MATRIX TECHNIQUES IN QUANTUM INFORMATION SCIENCE

CHI-KWONG LI

Department of Mathematics,
The College of William and Mary,
Williamsburg, Virginia 23187-8795, USA
E-mail: ckli@math.wm.edu

Mathematical techniques in quantum information science will be discussed. The focus will be on two specific topics. The first one concerns the study of quantum operations (channels) using the theory of completely positive linear maps. The second one concerns the study of quantum error correction using the theory of generalized numerical ranges. The discussion includes material from some recent research papers.

Keywords: Quantum operations, completely positive linear maps, quantum error corrections, higher rank numerical ranges.

1. Quantum Operations, Completely Positive Linear Maps

In this section, we describe the connections between quantum operations (channels) and completely positive linear maps. We present results of completely positive maps, and use them to study quantum operations which send certain quantum states to other specific quantum states.

1.1. *Open quantum systems*

A unitary time evolution of a closed system is determined by the quantum map $\mathcal{E}$ defined by

$$\mathcal{E}(\rho_S) = U(t)\rho_S U(t)^\dagger.$$

Here, ρ_S is the density matrix of a closed system at time $t = 0$ and $U(t)$ is the time evolution operator; for example, see [18, Section 2.1].

An open system is a system of interest (called the *principal system*) coupled with its environment. The total Hamiltonian is given by

$$H_T = H_S + H_E + H_{SE},$$

where H_S, H_E and H_{SE} are the system Hamiltonian, the environment Hamiltonian and their interaction Hamiltonian, respectively.

The state of the total system, which is assumed to be closed, will be described by ρ acting on the Hilbert space $\mathcal{H}_S \otimes \mathcal{H}_E$ such that one has the approximation $\rho(0) = \rho_S \otimes \rho_E$ and

$$\rho(t) = U(t)(\rho_S \otimes \rho_E)U(t)^\dagger \quad \text{for } t > 0.$$

For simplicity, we may assume that H_T is a constant matrix and $U(t) = e^{-itH_T}$. We study the principal system $(\mathcal{H}_S)$ by taking the *partial trace*

$$\begin{aligned}
\rho_S(t) &= \text{Tr}_E[U(t)(\rho_S \otimes \rho_E)U(t)^\dagger] \\
&= \sum_{a \in J}(I_S \otimes \langle \varepsilon_a|)[U(t)(\rho_S \otimes \rho_E)U(t)^\dagger](I_S \otimes |\varepsilon_a\rangle)
\end{aligned}$$

for any complete orthonormal basis $\{|\varepsilon_a\rangle : a \in J\}$ for $\mathcal{H}_E$, where J is an appropriate index set. We may assume that $\rho_E = |\varepsilon_0\rangle\langle\varepsilon_0|$ by linearity or by purification. Let $E_a(t) = (I_S \otimes \langle \varepsilon_a|)U(t)(I_S \otimes |\varepsilon_0\rangle)$. Then

$$\rho_S(t) = \sum_a E_a(t)\rho_S E_a(t)^\dagger.$$

This is known as the *operator-sum representation* of the quantum operation. Note that

$$\begin{aligned}
\sum_a E_a(t)^\dagger E_a(t) &= \sum_a (I_S \otimes \langle\varepsilon_0|)U(t)^\dagger(I_S \otimes |\varepsilon_a\rangle)(I_S \otimes \langle\varepsilon_a|)U(t)(I_S \otimes |\varepsilon_0\rangle) \\
&= (I_S \otimes \langle\varepsilon_0|)U(t)^\dagger(I_S \otimes I_E)U(t)(I_S \otimes |\varepsilon_0\rangle) = I_S.
\end{aligned}$$

This is the *trace preserving* condition for the quantum operation. For certain quantum operations or channels, one may relax this condition.

Remarks

- To understand quantum operations, one has to understand the set

$$\{E_a(t) : a \in J\}.$$

- How to detect and determine $\{E_a(t) : a \in J\}$ for a given quantum operation is a challenging problem. For example, one may study the images of different states ρ_S under the quantum operations to understand $\{E_a(t) : a \in J\}$. This is known as the *quantum process tomorgraphy*.

- The construction or simulation of quantum operations reduces to the construction or simulation of the set $\{E_a(t) : t \in J\}$.

Examples

- Noisy quantum channels are quantum operations. Sometimes, one may relax the trace preserving condition for quantum operations.
- Many quantum process such as measurement, positive operator-valued measure (POVM) can be viewed as quantum operations (channels); $E_a(t)$ is connected to the measurement operator M_a in this case.
- Suppose U_a is unitary, $p_a \in (0, 1]$, and $\sum_a p_a = 1$. A *mixing process* (a *mixed unitary channel*) is defined by

$$\mathcal{M}(\rho_S) = \sum_a p_a U_a \rho_S U_a^\dagger.$$

- In particular, we can let I_2 be the 2×2 identity matrix, and U_a be obtained from $I_2 \otimes \cdots \otimes I_2$ by changing one of the I_2 to one of the following Pauli matrices:

$$\sigma_x = \begin{bmatrix} 0 & 1 \\ 1 & 0 \end{bmatrix}, \quad \sigma_y = \begin{bmatrix} 0 & -i \\ i & 0 \end{bmatrix}, \quad \sigma_z = \begin{bmatrix} 1 & 0 \\ 0 & -1 \end{bmatrix}$$

to get a *Pauli channel*.

1.2. *Completely positive linear maps*

Definition Let r be a positive integer. A linear map $L : M_n \to M_m$ is *r-positive* if the map $I_r \otimes L : M_r(M_n) \to M_r(M_m)$ defined by

$$(I_r \otimes L)(A_{ij}) = [L(A_{ij})]$$

maps positive operators to positive operators. It is *completely positive* if it is r-positive for all positive integer r.

Let $\{|e_1\rangle, \ldots, |e_n\rangle\}$ be the standard basis for $\mathbf{C}^n$, and $E_{ij} = |e_i\rangle\langle e_j|$ for $1 \le i, j \le n$. Then $\{E_{11}, E_{12}, \ldots, E_{nn}\}$ the standard basis for M_n.

Example The map $L : M_n \to M_n$ defined by $L(A) = A^t$ is positive, but not 2-positive.

To see this, note that if $A = \begin{bmatrix} E_{11} & E_{12} \\ E_{21} & E_{22} \end{bmatrix}$, then $(I_2 \otimes L)(A) = \begin{bmatrix} E_{11} & E_{21} \\ E_{12} & E_{22} \end{bmatrix}$ has a principal matrix of the form $\begin{bmatrix} 0 & 1 \\ 1 & 0 \end{bmatrix}$, which is indefinite.

We have the following result by Choi[1,2]; see also[10].

Theorem 1.1. *Let* $\Phi : M_n \to M_m$. *The following are equivalent.*

(a) Φ *is completely positive.*
(b) Φ *is n-positive.*
(c) *The Choi matrix* $[\Phi(E_{ij})]$ *is positive semidefinite.*
(d) Φ *admits an operator-sum representation:*

$$\Phi(A) \mapsto \sum_{j=1}^{r} F_j A F_j^\dagger. \tag{$\ddagger$}$$

Proof. The implications (a) $\Rightarrow$ (b) $\Rightarrow$ (c) are clear. Suppose (c) holds. Let

$$[\Phi(E_{ij})] = [\,|f_1\rangle \cdots |f_r\rangle][\,|f_1\rangle \cdots |f_r\rangle]^\dagger = |f_1\rangle\langle f_1| + \cdots + |f_r\rangle\langle f_r|.$$

Arrange each $|f_j\rangle \in \mathbf{C}^{mn}$ as $m \times n$ matrix F_j by taking the first m entries as the first column, the next m entries as the second column, etc., for each j. One can check that

$$\Phi(E_{pq}) = \sum_{j=1}^{r} F_j E_{pq} F_j^\dagger.$$

By linearity, condition (d) holds.

Suppose (d) holds. Then

$$(I_k \otimes \Phi)(A_{ij}) = \sum_{j=1}^{r} (I_k \otimes F_j)(A_{ij})(I_k \otimes F_j)^\dagger$$

is a positive map for each k. Thus (a) holds. $\qquad\square$

Remarks concerning Theorem 1.1

- From mathematical point of view, checking the Choi matrix $[\Phi(E_{ij})]$ is the simplest and most natural thing to do, and it turns out to be the right test for complete positivity.
- The matrices F_j in ($\ddagger$) are called the *Choi operators* or the *Kraus operators*. In the context of quantum error correction, F_j are called the *error operators*.
- For any linear map $L : M_n \to M_m$ sending Hermitian matrices to Hermitian matrices, one can write $[L(E_{ij})] = XX^\dagger - YY^\dagger$, and show that $L(A) = \sum_{j=1}^{r} F_j A F_j^\dagger - \sum_{j=1}^{s} G_j A G_j^\dagger$.

Theorem 1.2. *Suppose* $\Phi : M_n \to M_m$ *of the form* (‡) *such that* $\{F_1, \ldots, F_r\}$ *is linearly independent, and* Φ *has another operator sum representation*

$$\Phi(A) = \sum_{j=1}^{s} \tilde{F}_j A \tilde{F}_j^{\dagger}.$$

Then there is $V = (v_{ij}) \in M_{r,s}$ *such that* $VV^* = I_r$ *and*

$$\tilde{F}_i = \sum_j v_{ij} F_j, \quad i = 1, \ldots, r.$$

Proof. Use the vector f_j corresponding to F_j to write

$$[\Phi(E_{ij})] = [\,|f_1\rangle \cdots |f_r\rangle][\,|f_1\rangle \cdots |f_r\rangle]^{\dagger}.$$

Also, use the vector f_j corresponding to $\tilde{F}_j$ to write

$$[\Phi(E_{ij})] = [\,|\tilde{f}_1\rangle \cdots |\tilde{f}_s\rangle][\,|\tilde{f}_1\rangle \cdots |\tilde{f}_r\rangle]^{\dagger}.$$

Since $[\,|f_1\rangle \cdots |f_r\rangle]$ has linearly independent columns spanning the column space of $[\Phi(E_{ij})]$, we see that

$$[\,|\tilde{f}_1\rangle \cdots |\tilde{f}_s\rangle] = [\,|f_1\rangle \cdots |f_r\rangle]V$$

for some $r \times s$ matrix $V = (v_{ij})$ such that $VV^* = I_r$. $\qquad\square$

Next, we present two basic results on completely positive linear maps. The proofs will be left as exercises.

Theorem 1.3. *Suppose* $\Phi : M_n \to M_m$ *is a completely positive linear map with operator sum representation* (‡). *Then*

$$\Phi(A) = F(I_r \otimes A)F^{\dagger} \quad \text{with } F = [F_1 \cdots F_r].$$

If $\tilde{F} = \begin{bmatrix} F_1 \\ \vdots \\ F_r \end{bmatrix}$ *and* $\tilde{F} A \tilde{F}^{\dagger} = (A_{ij})$ *with* $A_{11}, \ldots, A_{rr} \in M_m$, *then*

$$\Phi(A) = A_{11} + \cdots + A_{rr}.$$

(a) *The map* Φ *is unital if and only if* $FF^{\dagger} = \sum_{j=1}^{r} F_j F_j^{\dagger} = I_m$.

(b) *The map* Φ *is trace preserving if and only if* $\tilde{F}^{\dagger} \tilde{F} = \sum_{j=1}^{r} F_j^{\dagger} F_j = I_n$.

Denote by $\langle X, Y \rangle = \text{Tr}(XY^{\dagger})$ the usual inner product of two (rectangular) matrices X, Y of the same size.

Theorem 1.4. *Suppose* $\Phi : M_n \to M_m$ *is a completely positive linear map with operator sum representation in ($\ddagger$). Then the dual linear map* $\Phi^* : M_m \to M_n$ *defined by*

$$\Phi^*(B) = \sum_{j=1}^{r} F_j^{\dagger} B F_j$$

is the unique linear map satisfying $\langle \Phi(A), B \rangle = \langle A, \Phi^*(B) \rangle$ *for all* $(A, B) \in M_n \times M_m$. *Consequently,* Φ *is unital if and only if* Φ^* *is trace preserving.*

Theorem 1.5. *Let* $\Phi : M_n \to M_m$. *Then* Φ *is a completely positive linear map if and only if* Φ *is* r-*positive for* $r = \min\{m, n\}$.

Proof. (by Y.T. Poon) It suffices to prove that if $m < n$ and Φ is m-positive, then Φ is completely positive. This will follow from the following three assertions.

Assertion 1. Φ is positive if and only if Φ^* is positive.

Reason. Φ is positive if and only if $\Phi(|\xi\rangle\langle\xi|)$ is positive for any $|\xi\rangle \in \mathbf{C}^n$. It follows that $\langle\eta|\Phi(|\xi\rangle\langle\xi|)|\eta\rangle \geq 0$ for any $(\xi, \eta) \in \mathbf{C}^n \times \mathbf{C}^m$. Thus,

$$0 \leq \langle \Phi(|\xi\rangle\langle\xi|), |\eta\rangle\langle\eta| \rangle = \langle |\xi\rangle\langle\xi|, \Phi^*(|\eta\rangle\langle\eta|) \rangle.$$

Thus, $\Phi^*(|\eta\rangle\langle\eta|)$ is positive. Hence, Φ^* is positive.

Assertion 2. Φ is k-positive if and only if Φ^* is k-positive.

Reason. Similar to that of Assertion 1.

Assertion 3. Φ is completely positive if and only if Φ is completely positive.

Reason. By the dual map representation.

Since Φ^* is completely positive if and only if Φ^* is m-positive, the result follows. $\qquad\square$

Remarks

- Trace preserving completely positive maps correspond to quantum operations (channels) Φ described by the Schrödinger picture, in which the state of a system evolves with time.
- Unital completely positive maps correspond to the system in the Heisenberg picture, where Φ^* maps the observables of $\mathcal{H}_B$ to the observables of $\mathcal{H}_A$.

Exercises

- Prove Theorem 1.3.
- Prove Theorem 1.4.
- Provide the details of the proof of Theorem 1.5.

1.3. *Interpolating problems*

In this and the next subsection, we will focus on the following problem. The discussion is based on[12]; see also[17].

Problem 1.1. *Given $A_1, \ldots, A_k \in M_n$ and $B_1, \ldots, B_k \in M_m$, determine the necessary and sufficient condition for the existence of a completely positive linear map $\Phi : M_n \to M_m$ possibly with some special properties (e.g., $\Phi(I_n) = I_m$ or/and Φ is trace preserving) such that*

$$\Phi(A_j) = B_j \qquad for\ j = 1, \ldots, k.$$

Remarks concerning Problem 1.1

- Problem 1.1 can be viewed as an interpolation problem by completely positive maps.
- In the context of quantum information science, the study is related to whether certain quantum operations (channels) exist and hence can be constructed in the laboratories.
- It is also connected to quantum state tomorgraphy, i.e., known quantum states are used to probe a quantum process to find out how the process can be described.

Denote by H_n the set of Hermitian matrices in M_n. Since for a positive linear map Φ satisfying $\Phi(X) = Y$ if and only if $\Phi(X^*) = Y^*$ for any $(X, Y) \in M_n \times M_m$, we can focus on the study of Problem 1.1 for $\{A_1, \ldots, A_k\} \subseteq H_n$ and $\{B_1, \ldots, B_k\} \subseteq H_m$.

Definition A *nonnegative* matrix is *column* (respectively, *row*) *stochastic* if in each column (respectively, row) the entries sum up to 1. If A is both column and row stochastic, then (the necessarily square matrix) A is *doubly stochastic*.

Theorem 1.6. *Let $\{A_1, \ldots, A_k\} \subseteq H_n$ and $\{B_1, \ldots, B_k\} \subseteq H_m$ be two commuting families. Then there exist unitary matrices $U \in M_n$ and $V \in M_m$ such that $U^\dagger A_i U$ and $V B_i V^\dagger$ are diagonal matrices with diagonals $\mathbf{a}_i = (a_{i1}, \ldots, a_{in})$ and $\mathbf{b}_i = (b_{i1}, \ldots, b_{im})$ respectively, for $i = 1, \ldots, k$. The following conditions are equivalent.*

(a) *There is a completely positive map* $\Phi : M_n \to M_m$ *such that* $\Phi(A_i) = B_i$ *for* $i = 1, \dots, k$.

(b) *There is an* $n \times m$ *nonnegative matrix* $D = (d_{pq})$ *such that*

$$(b_{ij}) = (a_{ij})D.$$

Suppose (b) holds. For $1 \le j \le m$, *let* F_j *be the* $n \times n$ *matrix having the* j*th column equal to* $(\sqrt{d_{1j}}, \sqrt{d_{2j}}, \dots, \sqrt{d_{nj}})^t$ *and zero elsewhere. Then we have*

$$B_i = \sum_{j=1}^{r} (U F_j V)^\dagger A_i (U F_j V), \qquad i = 1, \dots, k. \tag{1}$$

Furthermore,

(i) Φ *in (a) is unital if and only if* D *in (b) can be chosen to be column stochastic.*

(ii) Φ *in (a) is trace preserving if and only if* D *in (b) can be chosen to be row stochastic.*

(iii) Φ *in (a) is unital and trace preserving if and only if* D *in (b) can be chosen to be doubly stochastic.*

Proof. We need the following observation for the proof of the theorem.

Suppose $\Phi : M_n \to M_m$ *is a completely positive map. Given unitaries* $U \in M_n$ *and* $V \in M_m$, *define* $\Psi : M_n \to M_m$ *by* $\Psi(X) = V^\dagger \Phi(U^\dagger X U) V$. *Then* Ψ *is also completely positive. Furthermore,* Ψ *is unital and/or trace preserving if and only if* Φ *has the corresponding property. If* $\Phi(X) = \sum_{j=1}^{r} F_j^\dagger X F_j$, *then* $\Psi(X) = \sum_{j=1}^{r} (U F_j V)^\dagger X (U F_j V)$.

By the above observation, we may assume that $A_i = \text{diag}(\mathbf{a}_i)$ and $B_i = \text{diag}(\mathbf{b}_i)$ for $i = 1, \dots, k$, and take $U = I_n$, $V = I_m$ in (1).

For $\mathbf{x} = (x_1, \dots, x_n)$, denote by $\text{diag}(\mathbf{x})$ the diagonal matrix with diagonal entries $x_1, \dots, x_n$.

(a) $\Rightarrow$ (b): Suppose there is a completely positive map $\Phi : M_n \to M_m$ such that $\Phi(A_i) = B_i$ for $i = 1, \dots, k$. By the operator sum representation in (‡), we have $F^j = [f_{pq}^j] \in M_{n,m}$, $j = 1, \dots, r$, such that

$$\Phi(X) = \sum_{j=1}^{r} F_j^\dagger X F_j.$$

For $1 \le p \le n$ and $1 \le q \le m$, let $d_{pq} = \sum_{j=1}^{r} |f_{pq}^j|^2$. Then $D = (d_{pq})$ is an $n \times m$ nonnegative matrix such that

$$(b_{ij}) = (a_{ij})D.$$

(b) $\Rightarrow$ (a): Suppose $D = (d_{pq})$ is a nonnegative matrix satisfying $(b_{ij}) = (a_{ij})D$. Let F_j be defined as in the theorem. Then direct computation shows that for every $\mathbf{x} = (x_1, \ldots, x_n)$, $\sum_{j=1}^{r} F_j^\dagger \text{diag}(\mathbf{x})F_j$ is a diagonal matrix with diagonal $\mathbf{y} = \mathbf{x}D$. Hence, for $i = 1, \ldots, k$, we have

$$\text{diag}(\mathbf{b}_i) = \sum_{j=1}^{m} F_j^\dagger \text{diag}(\mathbf{a}_i)F_j,$$

for $i = 1, \ldots, k$.

For $N \geq 1$, let $\mathbf{1}_N$ be a row vector of N 1's, then we have

Φ is unital $\Leftrightarrow \Phi(I_n) = I_m \Leftrightarrow \mathbf{1}_m = \mathbf{1}_n D \Leftrightarrow D$ is column stochastic.

This proves (i). The proof for cases (ii) and (iii) are similar. $\qquad\square$

Recall that a completely positive map $\Phi : M_n \to M_n$ is called *mixed unitary* if there exist unitary matrices $U_1, \ldots, U_r \in M_n$ and positive numbers $t_1, \ldots, t_r$ summing up to 1 such that

$$\Phi(X) = \sum_{j=1}^{r} t_j U_j^\dagger X U_j.$$

Clearly, every mixed unitary completely positive map is unital and trace preserving. For $n \geq 3$, there exists[15] a unital trace preserving completely positive map which is not mixed unitary.

By the Birkhoff Theorem[16], a doubly stochastic matrix D can be expressed in the form

$$D = \sum_{j=1}^{r} t_j P_j \qquad (2)$$

for some positive numbers $t_1, \ldots, t_r$ summing up to 1 and permutation matrices $P_1, \ldots, P_r \in M_n$. Using this result and (iii) in Theorem 1.6, we have the following corollary.

Corollary 1.1. *Under the hypothesis of Theorem 1.6, suppose there is a unital trace preserving completely positive map Φ satisfying (a), and D is a doubly stochastic matrix satisfying (b). Let D be expressed as in (2). Then the mixed unitary map $\Psi : M_n \to M_n$ defined by*

$$\Psi(X) = \sum_{j=1}^{r} t_j (UP_jV)^\dagger X (UP_jV)$$

also satisfies $\Psi(A_i) = B_i$ for $i = 1, \ldots, k$.

Remarks concerning Theorem 1.6

- Note that the same conclusion of Theorem 1.6 holds for completely positive maps without the unital/trace preserving requirement if the matrices U and V in the hypothesis are merely invertible instead of unitary. In other words, the result applies to two families $\{A_1, \ldots, A_k\} \subseteq H_n$ and $\{B_1, \ldots, B_k\} \in H_m$ such that each family is simultaneously congruent to diagonal matrices. For example, two Hermitian matrices A_1 and A_2 are simultaneously congruent to diagonal matrices if any one of the following conditions is satisfied.

  (i) $\alpha_1 A_1 + \alpha_2 A_2$ is positive definite.

  (ii) Both A_1 and A_2 are positive semidefinite.

- To check condition (b) and assertions (i) , (ii) and (iii), one can use standard linear programming techniques. In (i), one can divide the problem of finding a column stochastic D into m independent problems of finding nonnegative vectors $\mathbf{d}_q = (d_{1q}, \ldots, d_{nq})^t$ with entries summing up to one such that $(a_{ij})\mathbf{d}_q$ equals to the qth column of (b_{ij}) for $q = 1, \ldots, m$. Of course, the solution set of each of this problem is the convex polyhedron

$$
P_q = \left\{ (d_{1q}, \ldots, d_{nq})^t : d_{pq} \geq 0, \sum_{p=1}^{n} d_{pq} = 1, \right.
$$

$$
\left. \text{and } b_{iq} = \sum_{\ell=1}^{n} a_{i\ell} d_{\ell q}, i = 1, \ldots, k \right\}
$$

in $\mathbf{R}^n$. The extreme points of the polyhedron P_q, if non-empty, has at most k (respectively, $k + 1$) nonzero entries because we need n equalities among the inequality and equality constraints to determine an extreme point of P_q. Thus, at least $n - k$ of the inequality constraints $d_{pq} \geq 0$ have to be equalities to determine an extreme point. By the above discussion, one can construct a sparse matrix D as a solution using the extreme points in the solution sets P_q. Similarly, for (ii) we can construct an extreme point (with sparse pattern) of the set of row stochastic matrices D satisfying $(b_{ij}) = (a_{ij})D$. However, unlike (i), we cannot treat individual rows separately to reduce the complexity of the computation.

- In the construction of $F_1, \ldots, F_r$ in the last assertion of the theorem, we see that each F_j has rank at most $\ell = \min\{m, n\}$ so that the corresponding completely positive map Φ is a super ℓ-positive map[20].

- Note that for the diagonal matrix $X_i = U^\dagger A_i U$, the map $X_i \mapsto F_j^\dagger X_i F_j$ has only one nonzero entry at the (j,j) position obtained by taking a nonnegative combination of the diagonal entries of X_i. In the context of quantum information science, it is easy to implement the map (quantum operation) Φ, and all the actions only take place in the diagonal entries (classical channels) once $A_1, \ldots, A_k$ and $B_1, \ldots, B_k$ are in diagonal forms.

1.4. *Completely positive maps on a single matrix*

For a single matrix, we can give a more detailed analysis of the result in Theorem 1.6, and show that the study is related to other topics such as eigenvalue inequalities. Moreover, the results show that there are some results on trace preserving completely positive maps with no analogs for unital completely positive maps, and vice versa.

We first consider *unital completely positive maps*.

Theorem 1.7. *Suppose $A \in H_n$ and $B \in H_m$ have eigenvalues $a_1, \ldots, a_n$ and $b_1, \ldots, b_m$, respectively. Let $\mathbf{a} = (a_1, \ldots, a_n)$ and $\mathbf{b} = (b_1, \ldots, b_m)$. The following conditions are equivalent.*

(a) *There is a (unital) completely positive map $\Phi : M_n \to M_m$ such that $\Phi(A) = B$.*

(b) *There is a nonnegative (column stochastic) matrix $D = (d_{pq})$ such that $\mathbf{b} = \mathbf{a}D$.*

(c) *There are real numbers $\gamma_1, \gamma_2 \geq 0$ (with $\gamma_1 = \gamma_2 = 1$) such that*

$$\gamma_2 \min\{a_i : 1 \leq i \leq n\} \leq b_j \leq \gamma_1 \max\{a_i : 1 \leq i \leq n\}. \quad (3)$$

for all $1 \leq j \leq m$.

Proof. By Theorem 1.6, (a) and (b) are equivalent. Without loss of generality, we may assume that $a_1 \geq \cdots \geq a_n$ and $b_1 \geq \cdots \geq b_m$.

(b) $\Rightarrow$ (c): Suppose there is a nonnegative matrix $D = (d_{pq})$ such that $\mathbf{b} = \mathbf{a}D$. Let $\gamma_1 = \sum_{p=1}^n d_{p1}$ and $\gamma_2 = \sum_{p=1}^n d_{pm}$. Then for each $1 \leq j \leq m$, we have

$$\gamma_2 a_n = \sum_{p=1}^{n} d_{pm} a_n \leq \sum_{p=1}^{n} d_{pm} a_p = b_m$$

$$\leq b_j \leq b_1 = \sum_{p=1}^{n} d_{p1} a_p \leq \sum_{p=1}^{n} d_{p1} a_1 = \gamma_1 a_1.$$

If D is column stochastic, then it follows from definition that $\gamma_1 = \gamma_2 = 1$.

(c) $\Rightarrow$ (b): Suppose (c) holds. Then for each $i = 1, \ldots m$, there exists $0 \leq t_i \leq 1$ such that $b_i = t_i \gamma_1 a_1 + (1 - t_i) \gamma_2 a_n$. Let D be the $n \times m$ matrix

$$\begin{bmatrix} t_1 \gamma_1 & t_2 \gamma_1 & \cdots & t_m \gamma_1 \\ 0 & 0 & \cdots & 0 \\ \vdots & \vdots & \vdots & \vdots \\ 0 & 0 & \cdots & 0 \\ (1 - t_1)\gamma_2 & (1 - t_2)\gamma_2 & \cdots & (1 - t_m)\gamma_2 \end{bmatrix}.$$

Then $\mathbf{b} = \mathbf{a}D$ and D is column stochastic if $\gamma_1 = \gamma_2 = 1$. □

Corollary 1.2. *If $A \in H_n$ and $B \in H_m$ are nonzero positive semi-definite, then there is a completely positive map $\Phi : M_n \to M_m$ such that $\Phi(A) = B$, and there is a completely positive map $\Psi : M_m \to M_n$ such that $\Psi(B) = A$.*

Using Theorem 1.7, one can construct a pair of density matrices $(A, B) \in H_n \times H_m$ such that there is a unital completely positive map Φ such that $\Phi(A) = B$, but there is no unital completely positive map Ψ such that $\Psi(B) = A$.

Next, we turn to *trace preserving completely positive maps*.

Theorem 1.8. *Suppose $A \in H_n$ and $B \in H_m$ have eigenvalues $a_1, \ldots, a_n$ and $b_1, \ldots, b_m$ respectively. Let $\mathbf{a} = (a_1, \ldots, a_n)$ and $\mathbf{b} = (b_1, \ldots, b_m)$. The following conditions are equivalent.*

(a) *There is a trace preserving completely positive map $\Phi : M_n \to M_m$ such that $\Phi(A) = B$.*

(b) *There exists an $n \times m$ row stochastic matrix D such that $\mathbf{b} = \mathbf{a}D$. Moreover, we can assume that the pth and qth row are identical whenever $a_p a_q > 0$, and the pth row of D can be arbitrary (non-negative with entries summing up to 1) if $a_p = 0$.*

(c) *We have $\mathrm{Tr} A = \mathrm{Tr} B$ and $\sum_{p=1}^{n} |a_p| \geq \sum_{q=1}^{m} |b_q|$. Equivalently, $\mathrm{Tr} A = \mathrm{Tr} B$ and the sum of the positive (negative) eigenvalues of A*

is not smaller (not larger) than the sum of the positive (negative) eigenvalues of B.

Proof. For simplicity, we assume that $a_1 \geq \cdots a_r \geq 0 > a_{r+1} \geq \cdots \geq a_n$ and $b_1 \geq \cdots \geq b_s \geq 0 > b_{s+1} \geq \cdots \geq b_m$. Let $a_+ = \sum_{p=1}^{r} a_p$, $a_- = \sum_{p=r+1}^{n} a_p$ and $b_+ = \sum_{q=1}^{s} b_q$, $b_- = \sum_{q=s+1}^{m} b_q$.

By Theorem 1.6, we have (b) $\Rightarrow$ (a). Also, by Theorem 1.6, if (a) holds, then $\mathbf{b} = \mathbf{a}D$ for an $n \times m$ row stochastic matrix D. Next, we show that D can be chosen to satisfy the second assertion of condition (b). To this end, let $\mathbf{d}_1, \ldots, \mathbf{d}_n$ be the rows of D. Clearly, if $a_p = 0$, we can replace the pth row of D by any nonnegative vectors with entries summing up to 1 to get $\tilde{D}$ and we still have $\mathbf{a}\tilde{D} = \mathbf{a}D = \mathbf{b}$. Now, suppose $a_1 > 0$. Then $a_+ > 0$. We can replace the first r rows (or the rows correspond to $a_p > 0$) by

$$\mathbf{d}_+ = \sum_{j=1}^{r} \frac{a_j}{a_+} \mathbf{d}_j$$

to obtain $\tilde{D}$. Then $\mathbf{d}_+$ has nonnegative entries summing up to 1, and $\mathbf{a}\tilde{D} = \mathbf{a}D = \mathbf{b}$. Similarly, suppose $a_n < 0$. Then $a_- < 0$. We can further replace the last $n - s$ rows of D by

$$\mathbf{d}_- = \sum_{j=s+1}^{n} \frac{a_j}{a_-} \mathbf{d}_j$$

to obtain $\tilde{D}$. Then $\mathbf{d}_-$ has nonzero entries summing up to 1, and $\mathbf{a}\tilde{D} = \mathbf{a}D = \mathbf{b}$.

(b) $\Rightarrow$ (c): Suppose $\mathbf{b} = \mathbf{a}D$. We have

$$\mathrm{Tr}B = \sum_{q=1}^{m} b_q = \sum_{q=1}^{m} \sum_{p=1}^{n} a_p d_{pq} = \sum_{p=1}^{n} a_p \left(\sum_{q=1}^{m} d_{pq} \right) = \sum_{p=1}^{n} a_p = \mathrm{Tr}A$$

and

$$\sum_{q=1}^{m} |b_q| = \sum_{q=1}^{m} \left| \sum_{p=1}^{n} a_p d_{pq} \right| \leq \sum_{q=1}^{m} \sum_{p=1}^{n} |a_p| d_{pq} = \sum_{p=1}^{n} |a_p| \left(\sum_{q=1}^{m} d_{pq} \right) = \sum_{p=1}^{n} |a_p|.$$

Since

$$\sum_{i=1}^{n} a_i = \sum_{j=1}^{m} b_j, \quad \sum_{i=1}^{n} |a_i| = a_+ - a_- = \sum_{i=1}^{n} a_i - 2a_- = 2a_+ - \sum_{i=1}^{n} a_i,$$

and

$$\sum_{j=1}^{m} |b_j| = b_+ - b_- = \sum_{j=1}^{m} b_j - 2b_- = 2b_+ - \sum_{j=1}^{m} b_j,$$

the last assertion of (c) follows.

(c) $\Rightarrow$ (b): Suppose $\mathrm{Tr}A = \mathrm{Tr}B$, $a_+ \geq b_+$ and $a_- \leq b_-$. Let

$$
t_q = \begin{cases} \dfrac{b_q}{a_+} & \text{for } 1 \leq q \leq s, \\[3mm] \dfrac{b_q}{a_-} & \text{for } s < q \leq m. \end{cases}
$$

Here, if $a_+ = 0$ then $b_+ = 0$, and we can set $t_q = 0$ for $1 \leq q \leq s$. If $a_- = 0$ then $b_- = 0$, and $s = m$. Therefore, $t_q \geq 0$ for all $1 \leq q \leq m$. We have

$$
a_+ \geq b_+ = (a_+)\sum_{q=1}^{s} t_q, \quad \text{and} \quad |a_-| \geq |b_-| = |a_-|\sum_{q=s+1}^{n} t_q.
$$

Let $u = 1 - \sum_{q=1}^{s} t_q \geq 0$, $v = 1 - \sum_{q=s+1}^{n} t_q \geq 0$ and D be an $n \times m$ row stochastic matrix with

$$
p\text{th row} = \begin{cases} (t_1, t_2, \ldots, t_s, 0, \ldots, 0, u) & \text{for } 1 \leq p \leq r, \\[3mm] (0, \ldots, 0, t_{s+1}, t_{s+2}, \ldots, t_{m-1}, t_m + v) & \text{for } r+1 \leq p \leq n. \end{cases}
$$

Since

$$
ua_+ + va_- = (a_+ - b_+) + (a_- - b_-) = 0,
$$

we have $\mathbf{b} = \mathbf{a}D$. $\qquad\square$

Corollary 1.3. *If $A \in H_n$ and $B \in H_m$ are density matrices, i.e. positive semi-definite and $\mathrm{Tr}A = \mathrm{Tr}B = 1$, then there is a completely positive map $\Phi : M_n \to M_m$ such that $\Phi(A) = B$, and there is a completely positive map $\Psi : M_m \to M_n$ such that $\Psi(B) = A$.*

As mentioned after Corollary 1.2, one may not be able to find a unital completely positive map taking a density matrix $B \in H_m$ to another density matrix $A \in H_n$ even if there is a unital positive completely positive map sending A to B.

Finally, we consider *unital trace preserving completely positive maps*.

Suppose there is a unital completely positive map sending A to B, and also a trace preserving completely positive map sending A to B. Is there a unital trace preserving completely positive map sending A to B? The following example shows that the answer is negative.

Example Suppose $A = \mathrm{diag}\,(4, 1, 1, 0)$ and $B = \mathrm{diag}\,(3, 3, 0, 0)$. By Theorems 1.7 and 1.8 there is a trace preserving completely positive map sending A to B, and also a unital completely positive map sending A to B. Let

$A_1 = A - I_4 = \mathrm{diag}\,(3,0,0,-1)$ and $B_1 = B - I_4 = \mathrm{diag}\,(2,2,-1,-1)$. By Theorem 1.8, there is no trace preserving completely positive map sending A_1 to B_1. Hence, there is no unital trace preserving completely positive map sending A to B.

As shown in Corollary 1.1, if there is a unital trace preserving map sending a commuting family in H_n to a commuting family in H_m, then we may chose the map to be mixed unitary. In the following, we show that for the case when $k = 1$ in Theorem 1.6, one can even assume that the map is the average of n unitary similarity transforms.

Let $\mathbf{a} = (a_1, \ldots, a_n), \mathbf{b} = (b_1, \ldots, b_n) \in \mathbf{R}^n$. We say that $\mathbf{b}$ is majorized by $\mathbf{a}$ ($\mathbf{b} \prec \mathbf{a}$) if for every $1 \le k < n$, the sum of the k largest entries of $\mathbf{b}$ is less than or equal to the sum of the k largest entries of $\mathbf{a}$, and $\sum_{i=1}^{n} a_i = \sum_{i=1}^{n} b_i$.

Theorem 1.9. *Suppose $A \in H_n$ and $B \in H_m$ have eigenvalues $a_1, \ldots, a_n$ and $b_1, \ldots, b_m$ respectively. Let $\mathbf{a} = (a_1, \ldots, a_n)$ and $\mathbf{b} = (b_1, \ldots, b_m)$. The following are equivalent.*

- *(a) There exists a unital trace preserving completely positive map Φ such that $\Phi(A) = B$.*
- *(a1) There exists a mixed unitary completely positive map Φ such that $\Phi(A) = B$.*
- *(a2) There exist unitary matrices $U_1, \ldots, U_n \in M_n$ such that $B = \frac{1}{n} \sum_{j=1}^{n} U_j^{\dagger} A U_j$.*
- *(a3) For each $t \in \mathbf{R}$, there exists a trace preserving completely positive map Φ_t such that $\Phi_t(A - tI) = B - tI$.*
- *(b) There is a doubly stochastic matrix D such that $\mathbf{b} = \mathbf{a}D$.*
- *(b1) There is a unitary matrix $W \in M_n$ such that $W\,\mathrm{diag}\,(a_1, \ldots, a_n)W^{\dagger}$ has diagonal entries $b_1, \ldots, b_n$.*
- *(c) $\mathbf{b} \prec \mathbf{a}$.*

Moreover, if condition (b) holds and $D = \sum_{\ell=1}^{r} t_\ell P_\ell$ such that $t_1, \ldots, t_r$ are positive numbers summing up to 1 and $P_1, \ldots, P_r$ are permutation matrices, then $B = \sum_{j=1}^{r} V^{\dagger} P_j^t U^{\dagger} A U P_j V$, where $U^{\dagger} A U = \mathrm{diag}\,(a_1, \ldots, a_n)$ and $V^{\dagger} B V = \mathrm{diag}\,(b_1, \ldots, b_n)$.

Proof. (b) $\Longleftrightarrow$ (c) is a standard result of majorization; see[16].

The implications (a2) $\Rightarrow$ (a1) $\Rightarrow$ (a) $\Rightarrow$ (a3) are obvious.

(a3) $\Rightarrow$ (c) : We may assume that $a_1 \ge \cdots \ge a_n$ and $b_1 \ge \cdots \ge b_n$. For $1 \le k < n$ choose t such that $a_k \ge t \ge a_{k+1}$. Then there is a trace preserving completely positive linear map Φ_t such that $\Phi_t(A - tI_n) = B - tI_n$. By

Theorem 1.8, the sum of the k positive eigenvalues of $B - tI_n$ is no larger than that of $A - tI_n$. Thus,

$$\sum_{i=1}^{k} b_i - kt = \sum_{i=1}^{k}(b_i - t) \leq \sum_{i=1}^{k}(a_i - t) = \sum_{i=1}^{k} a_i - kt. \qquad (4)$$

We see that $\sum_{i=1}^{k} b_i \leq \sum_{i=1}^{k} a_i$ for $k = 1, \ldots, n-1$. Since $\mathrm{Tr} A = \mathrm{Tr} B$, we have $\mathbf{b} \prec \mathbf{a}$.

(c) $\Rightarrow$ (b1) is a result of Horn[8].

(b1) $\Rightarrow$ (a2) : Suppose W is a unitary matrix such that the diagonal of $W \mathrm{diag}\,(a_1, \ldots, a_n) W^\dagger$ has diagonal entries $b_1, \ldots, b_n$. Let $w = e^{i2\pi/n}$, $P = \mathrm{diag}\,(1, w, \ldots, w^{n-1})$. Then

$$B = \frac{1}{n} \sum_{j=1}^{n} UW(P^j)^\dagger WUAU^\dagger W^\dagger (P^j) U^\dagger.$$

Thus, (a2) holds. $\qquad \square$

In the context of quantum information theory, a completely positive map in condition (a1) of the above theorem is a *mixed unitary quantum channel/operation*. By the above theorem, the existence of a mixed unitary quantum channel Φ taking a quantum state $A \in H_n$ to a quantum state $B \in H_m$ can be described in terms of trace preserving completely positive maps, namely, condition (a3). However, despite the duality of the two classes of maps, there is no analogous condition in terms of unital completely positive map; see Proposition 1.4.

Remark

- For non-commuting family, Problem 1.1 is very challenging and is still under current research. One needs to use other tools such as dilation, numerical range, etc. to study the problem.
- Also, there are studies of Problem 1.1 with the extra restriction that Φ has a bounded Choi/Kraus rank. This corresponds to the problem of constructing the quantum operations with the minimum number of Choi/Kraus operators $F_1, \ldots, F_r$.

Exercises

- Suppose $x = (x_1, \ldots, x_n)$ has nonnegative entries summing up to 1. Show that

$$(1, \ldots, 1)/n \prec x \prec (1, 0, \ldots, 0).$$

From this result, deduce Corollary 1.3.

- Give an example of density matrices $A \in M_n$ and $B \in M_m$ such that there is no unital completely positive linear map sending A to B or sending B to A.

2. Quantum Error Correction, Higher Rank Numerical Ranges

In this section, we present an operator theory approach to quantum error correction. In particular, numerical range techniques in the study will be presented.

2.1. *Algebraic approach to quantum error correction*

In classical coding theory, to transmit a binary bit information $x \in \{0, 1\}$ one may encode x by xxx. If there is a probability p for a bit-flip, then the received messages may take the following forms and decoded correctly as xxx or yyy, where $y = x \oplus 1$, by the majority vote scheme as follows.

Received message	Decoded message	Probability
xxx	xxx	$(1-p)^3$
yxx	xxx	$(1-p)^2 p$
xyx	xxx	$(1-p)^2 p$
xxy	xxx	$(1-p)^2 p$
yyx	yyy	$(1-p)p^2$
yxy	yyy	$(1-p)p^2$
xyy	yyy	$(1-p)p^2$
yyy	yyy	p^3

The probability of correct decoding is

$$p_0 = (1-p)^3 + 3p(1-p)^2 = (1-p)^2(1+2p),$$

which is much larger than the probability of failure: $(3-2p)p^2$ if p is small.

Three-Qubit Bit-Flip QECC

Assume that there is a probability of p for bit-flip so that

$$|\psi\rangle = a|0\rangle + b|1\rangle \mapsto X|\psi\rangle = a|1\rangle + b|0\rangle.$$

Here we derive a QECC under the constraints of no-cloning, and that measurement of a quantum state will collapse it.

Step 1 Encoding $|\psi\rangle|00\rangle \mapsto |\psi\rangle_L = a|000\rangle + b|111\rangle$.

Step 2 Transmission

Step 3 Error syndrome detection and correction

From the receiving state $|x_1x_2x_3\rangle$, construct the syndrome

$$|AB\rangle = |x_1 \oplus x_2\rangle|x_1 \oplus x_3\rangle.$$

We then apply one of the corrections.

(a) If $|AB\rangle = |00\rangle$, do nothing.
(b) If $|AB\rangle = |11\rangle$, apply $X_1 = X \otimes I \otimes I$.
(c) If $|AB\rangle = |10\rangle$, apply $X_2 = I \otimes X \otimes I$.
(d) If $|AB\rangle = |01\rangle$, apply $X_3 = I \otimes I \otimes X$.

Received state	Syndrome	Decoded state	Probability
$a|000\rangle + b|111\rangle$	$|00\rangle$	$a|000\rangle + b|111\rangle$	$(1-p)^3$
$a|100\rangle + b|011\rangle$	$|11\rangle$	$a|000\rangle + b|111\rangle$	$(1-p)^2 p$
$a|010\rangle + b|101\rangle$	$|10\rangle$	$a|000\rangle + b|111\rangle$	$(1-p)^2 p$
$a|001\rangle + b|110\rangle$	$|01\rangle$	$a|000\rangle + b|111\rangle$	$(1-p)^2 p$
$a|110\rangle + b|001\rangle$	$|01\rangle$	$a|111\rangle + b|000\rangle$	$(1-p)p^2$
$a|101\rangle + b|010\rangle$	$|10\rangle$	$a|111\rangle + b|000\rangle$	$(1-p)p^2$
$a|011\rangle + b|100\rangle$	$|11\rangle$	$a|111\rangle + b|000\rangle$	$(1-p)p^2$
$a|111\rangle + b|000\rangle$	$|00\rangle$	$a|111\rangle + b|000\rangle$	p^3

The probability of error will be $p^2(3 - 2p)$, which is small.

Hamming Code and constructing QECC by algebraic method

Classical Theory

Construction of $(7, 4, 3)$ Hamming code.

Suppose messages represented by binary sequences v of length four.

Step 1 Construct a 4×7 zero-one matrix M so that each v is transformed into vM for transmission so that each pair of code words have a Hamming distance at least 3.

Step 2 Assume that no more than one error in the transmission with a high probability.

Step 3 Construct a 3×7 parity check matrix H. For the received word c, compute the syndrome by $Hc^t = (x_2, x_1, x_0)^t$ to indicate that a bit-flip occurs at the $x = x_2 x_1 x_0$ position (with a high probability). Then do the decoding accordingly.

Extending this idea, Steane, Claderbank and Shor developed the **seven-qubit QECC**. This improved the nine-qubit QECC by Shor. Later, an "optimal" five-qubit QECC was developed to correct errors on a single qubit under the assumption that a bit-flip or phase-flip acts independently on one of the five qubits.

2.2. *Operator approach to quantum error correction*

A quantum channel can be viewed as a trace preserving completely positive linear map $\Phi : M_n \to M_m$ of the form

$$\Phi(A) \mapsto \sum_{j=1}^{r} F_j A F_j^\dagger. \qquad (\ddagger)$$

Let $\mathbf{V}$ be a subspace of $\mathbf{C}^n$ and $P_{\mathbf{V}}$ the orthogonal projection of $\mathbf{C}^n$ onto $\mathbf{V}$. Then $\mathbf{V}$ is a quantum error correction code for Φ if there exists a quantum operation (trace preserving completely positive linear map) $\Psi : M_n \to M_n$ such that $\Psi \circ \Phi(A) = A$ for all $A \in P_{\mathbf{V}} M_n P_{\mathbf{V}}$. We have the following; see[6,9].

Theorem 2.1. *Let* $\Phi : M_n \to M_n$ *be a quantum channel of the form* $(\ddagger)$. *Suppose* $\mathbf{V} \subseteq \mathbf{C}^n$ *is a subspace and* $P_{\mathbf{V}} \subseteq M_n$ *is the orthogonal projection with* $\mathbf{V}$ *as the range space. Then* $\mathbf{V}$ *is an quantum error correction if and only if there are scalars* γ_{ij} *with* $1 \leq i, j \leq m$ *such that*

$$P_{\mathbf{V}} F_i^\dagger F_j P_{\mathbf{V}} = \gamma_{ij} P_{\mathbf{V}}, \quad 1 \leq i, j \leq m.$$

Proof. See [19, Section 10.3]. □

Example Consider the three qubit bit-flip channel $\Phi : M_8 \to M_8$ with error operators $\sqrt{p_1} X_1$, $\sqrt{p_2} X_2$, $\sqrt{p_3} X_3$, and $\sqrt{1 - p_1 - p_2 - p_3} I$. Then $V = \text{span} \{|000\rangle, |111\rangle\}$ is a QECC. The recovery channel is defined by

$$\Psi(A) = A + P(X_1^\dagger A X_1 + X_2^\dagger A X_2 + X_3^\dagger A X_3)P.$$

Exercise Verify that $\Psi \circ \Phi(A) = A$ if $PAP = P$.
Hint: It suffices to check the equality for $A = |x\rangle\langle x|$ for $|x\rangle$ in the range space of P. In fact, since P has rank two, one needs only to check the equality for three linearly independent rank-one orthogonal projections.

2.3. *Higher rank numerical ranges and basic properties*

Motivated by Theorem 2.1, researchers study the *(joint) higher rank numerical range* $\Lambda_k(\mathbf{A})$ of the matrices $\mathbf{A} = (A_1, \ldots, A_m)$ defined as the collection of $(a_1, \ldots, a_m) \in \mathbf{C}^{1 \times m}$ such that

$$PA_j P = a_j P, \qquad j = 1, \ldots, m,$$

for some rank-k orthogonal projection P.

Exercise Show that $(a_1, \ldots, a_m) \in \Lambda_k(A_1, \ldots, A_m)$ if and only if any one of the following holds.

(a) There is a unitary $U \in M_n$ such that the leading $k \times k$ principal submatrix of $U^\dagger A_j U$ equals $a_j I_k$ for $j = 1, \ldots, m$.

(b) There is an $n \times k$ matrix X such that $X^\dagger X = I_k$ and

$$X^\dagger A_j X = a_j I_k, \quad j = 1, \ldots, m.$$

Note that $\Lambda_1(A_1, \ldots, A_m)$ reduces to the (classical) joint numerical range of the matrices $A_1, \ldots, A_m$ defined and denoted by

$$W(A_1, \ldots, A_m) = \{(\langle x|A_1|x\rangle, \ldots, \langle x|A_m|x\rangle) : |x\rangle \in \mathbf{C}^n, \langle x|x\rangle = 1\},$$

which is useful in the study of matrices and operators; see[7].

It turns out that even for a single matrix A, the rank-k numerical range $\Lambda_k(A)$ has non-trivial properties. Denote by $\lambda_k(H)$ the k-th largest eigenvalue of the Hermitian matrix $H \in M_n$. We have the following.

Theorem 2.2. *Let $A \in M_n$ and $k \in \{1, \ldots, n-1\}$.*

P1. For any $a, b \in \mathbf{C}$, $\Lambda_k(aA + bI) = a\Lambda_k(A) + b$.

P2. For any unitary $U \in M_n$, $\Lambda_k(U^\dagger AU) = \Lambda_k(A)$.

P3. For any $n \times r$ matrix V with $r \geq k$ and $V^\dagger V = I_r$, we have

$$\Lambda_k(V^\dagger AV) \subseteq \Lambda_k(A).$$

P4. If A is Hermitian, then $\Lambda_k(A) = [\lambda_{n-k+1}(A), \lambda_k(A)]$, where the interval is an empty set if $\lambda_{n-k+1}(A) > \lambda_k(A)$ when $k > n/2$.

P5. If $\mu \in \Lambda_k(A)$, then for any $\xi \in [0, 2\pi)$,

$$e^{i\xi}\mu + e^{-i\xi}\bar{\mu} \leq \lambda_k(e^{i\xi}A + e^{-i\xi}A^\dagger).$$

P6. Suppose $n < 2k$. The set $\Lambda_k(A)$ has at most one element.

Exercise Verify Theorem 2.2.

Hint: The verification of (P1) – (P3) is straight forward.

To verify (P4), use the generalized inequalities to deduce that if $UAU^\dagger$ has the leading principal submatrix μI_k then $\lambda_k(A) \geq \mu \geq \lambda_{n-k+1}(A)$; conversely, if $\mu \in [\lambda_{n-k+1}(A), \lambda_k(A)]$ then one can construct U such that $UAU^\dagger$ has leading principal submatrix μI_k.

(P5) and (P6) can be readily deduced from (P4).

Theorem 2.3. *Let $A \in M_n$ and $k \in \{1, \ldots, n\}$.*

(a) If $n \geq 3k - 2$, then $\Lambda_k(A)$ is non-empty.

(b) If $n < 3k - 2$, there is $B \in M_n$ such that $\Lambda_k(B) = \emptyset$.

(c) We have the following equality

$$\Lambda_k(A) = \bigcap_{\xi \in [0, 2\pi)} \left\{ \mu \in \mathbf{C} : e^{i\xi}\mu + e^{-i\xi}\bar{\mu} \leq \lambda_k(e^{i\xi}A + e^{-i\xi}A^\dagger) \right\}.$$

(d) $\Lambda_k(A)$ is always convex.

(e) If $A \in M_n$ is a normal matrix with eigenvalues $\lambda_1, \ldots, \lambda_n$, then

$$\Lambda_k(A) = \bigcap_{1 \leq j_1 < \cdots < j_{n-k+1} \leq n} \text{conv}\{\lambda_{j_1}, \ldots, \lambda_{j_{n-k+1}}\}.$$

Proof. For (a) and (b), see[13] and also[3]. For (c) – (e), see[15]. $\qquad\square$

Example (Bi-unitary channels.) Suppose $U_1, U_2 \in M_n$ are unitary, $p \in (0, 1)$, and

$$\Phi(X) = pU_1 X U_1^\dagger + (1 - p)U_2 X U_2^\dagger.$$

Then Φ has a QECC of dimension k if and only if $\Lambda_k(U_1^\dagger U_2) \neq \emptyset$ because $U_1^\dagger U_1 = U_2^\dagger U_2 = I_n$ and $(U_1^\dagger U_2)^\dagger = U_2^\dagger U_1$.

Exercise For a bi-unitary channel, if $n = 4$ and the unitary matrix $A = U_1^\dagger U_2 \in M_4$ has eigenvalues $\mu_1, \ldots, \mu_4$, show that there is a QECC of dimension 2 because one of the following holds.

(a) If $\text{conv}\{\mu_1, \mu_2, \mu_3, \mu_4\}$ has four vertices. then $\Lambda_2(A) = \{\mu\}$, where μ is the intersection of the two diagonals of the quadrilateral.

(b) If $\text{conv}\{\mu_1, \mu_2, \mu_3, \mu_4\}$ has three vertices μ_1, μ_2, μ_3, then $\Lambda_2(A) = \{\mu_4\}$.

(c) If $\text{conv}\{\mu_1, \mu_2, \mu_3, \mu_4\}$ is a (possibly degenerate) line segment with vertices μ_1, μ_2, then $\Lambda_2(A)$ is a line segment joining μ_3 and μ_4.

Example If $A \in M_8$ has eigenvalues $1, w, \ldots, w^7$ with $w = e^{i2\pi/8}$, then

$$\Lambda_2(A) = \text{conv}\{w, -w, iw, -iw\} \cap \text{conv}\{1, -1, i, -i\}, \quad \text{and} \quad \Lambda_4(A) = \{0\}.$$

Remarks

- In the above example, if we change A to A' so that $X = A' - A$ has a small norm, then most likely we have $\Lambda_4(A') = \emptyset$. On the other hand, if $X = A' - A$ has low rank, there are better perturbation results.
- Actually, one can determine $\Lambda_k(A)$ by intersecting $\Lambda_{k-r}(A+X)$ for all rank-r matrix X (of a certain type) as shown in the following result.

Theorem 2.4. *Suppose* $A \in M_n$ *and* $1 \le r < k \le n$. *Let* $\mathcal{S} \subseteq \{X \in M_n : \text{rank}(X) \le r\}$ *contain all matrices of the form* μP *with* $|\mu| = 2\|A\|$ *and* $P \in \mathcal{P}_r$, *where* $\mathcal{P}_r$ *is the set of rank-r orthogonal projections in* M_n. *Then*

  (a) $\Lambda_k(A) = \cap\{\Lambda_{k-r}(A + F) : F \in \mathcal{S}\}$.
  (b) $\Lambda_k(A) = \cap\{\Lambda_{k-1}(A + 2e^{i\xi}\|A\|P) : \xi \in [0, 2\pi), P \in \mathcal{P}_1\}$.
  (c) $\Lambda_k(A) = \cap\{W(A + 2e^{i\xi}\|A\|P) : \xi \in [0, 2\pi), \ P \in \mathcal{P}_{k-1}\}$.

Proof. The proof can be found in[14]. $\qquad\qquad\qquad\qquad\qquad\qquad\qquad$ $\square$

2.4. *Results on the joint higher rank numerical range*

In this subsection, we present results on the joint higher rank numerical range of several matrices. The proofs can be found in[11].

Suppose

$$A_j = H_{2j-1} + iH_{2j} \quad \text{with} \quad H_{2j-1}, H_{2j} \in H_n \quad \text{for } j = 1, \ldots, m.$$

Then $\Lambda_k(\mathbf{A}) \subseteq \mathbf{C}^m$ can be identified with $\Lambda_k(H_1, \ldots, H_{2m}) \subseteq \mathbf{R}^{2m}$. Thus, we will focus on the joint rank-k numerical ranges of self-adjoint operators in our discussion.

Here are some basic properties.

Proposition 2.1. *Suppose* $\mathbf{A} = (A_1, \ldots, A_m) \in H_n^m$, *and* $T = (t_{ij})$ *is an* $m \times r$ *real matrix. If* $B_j = \sum_{i=1}^{m} t_{ij} A_i$ *for* $j = 1, \ldots, r$, *then*

$$\{(a_1, \ldots, a_m)T : (a_1, \ldots, a_m) \in \Lambda_k(\mathbf{A})\} \subseteq \Lambda_k(\mathbf{B}).$$

Equality holds if $\{A_1, \ldots, A_m\}$ *is linearly independent and*

$$\text{span}\{A_1, \ldots, A_m\} = \text{span}\{B_1, \ldots, B_r\}.$$

In view of the above proposition, in the study of the geometric properties of $\Lambda_k(\mathbf{A})$, we may always assume that $A_1, \ldots, A_m$ are linearly independent.

Proposition 2.2. *Let* $\mathbf{A} = (A_1, \ldots, A_m) \in H_n^m$, *and let* $k < n$.
(a) For any real vector $\mu = (\mu_1, \ldots, \mu_m)$,

$$\Lambda_k(A_1 - \mu_1 I, \ldots, A_m - \mu_m I) = \Lambda_k(\mathbf{A}) - \mu.$$

(b) If $(a_1, \ldots, a_m) \in \Lambda_k(\mathbf{A})$ *then* $(a_1, \ldots, a_{m-1}) \in \Lambda_k(A_1, \ldots, A_{m-1})$.
(c) $\Lambda_{k+1}(\mathbf{A}) \subseteq \Lambda_k(\mathbf{A})$.

Remark By Proposition 2.2 (a), we can replace A_j by $A_j - \mu_j I$ for $j = 1, \ldots, m$, without affecting the geometric properties of $\Lambda_k(A_1, \ldots, A_m)$.

Theorem 2.5. *Let* $\mathbf{A} \in H_n^m$. *For* $m \geq 1$ *and* $k > 1$. *Then* $\Lambda_k(\mathbf{A})$ *is non-empty if*

$$n \geq (k-1)(m+1)^2.$$

If $\Lambda_k(A_1, \ldots, A_m)$ has a non-trivial convex subset, then it is more resistant to perturbation. Recall that a set $\mathcal{S} \subseteq \mathbf{R}^m$ is star shaped with a star center μ_0 is for every $\mu \in \mathcal{S}$, the line segment joining μ_0 and μ lies entirely in $\mathcal{S}$.

Theorem 2.6. *Let* $\mathbf{A} = (A_1, \ldots, A_m) \in H_n^m$ *and* $k \in \{1, \ldots, n-1\}$. *If* $\Lambda_{\hat{k}}(\mathbf{A}) \neq \emptyset$ *for some* $\hat{k} \geq (m+2)k$, *then* $\Lambda_k(\mathbf{A})$ *is star-shaped and contains the convex subset* $\text{conv}\Lambda_{\hat{k}}(\mathbf{A})$ *so that every element in* $\text{conv}\Lambda_{\hat{k}}(\mathbf{A})$ *is a star center of* $\Lambda_k(\mathbf{A})$.

Remark There are many open problems for joint higher rank numerical range.

- If A_1, A_2, A_3 mutually commute and $n \geq 5$, then $\Lambda_2(A_1, A_2, A_3)$ is non-empty. For general $A_1, A_2, A_3 \in H_n$, we only know that $\Lambda_2(A_1, A_2, A_3)$ is non-empty if $n \geq 7$. How about $n = 5, 6$?
- It would be interesting to determine the minimum n (depending on m and k) for which $\Lambda_k(A_1, \ldots, A_m)$ is non-empty, star-shaped, or convex.

Exercise Prove Propositions 2.5 and 2.6.

Acknowledgment

The author would like to thank the organizer of the Summer School on Diversities in Quantum Computation/Information for their excellent organization work and warm hospitality. The support of the Research Center of Quantum Computing at the Kinki University, and the support of USA NSF are graciously acknowledged. Furthermore, he would like to thank Professor Mikio Nakahara, Akira SaiToh, Utkan Gungordu for the many helpful comments to the lecture notes.

References

1. M.D. Choi, Positive linear maps on C^*-algebras, Canad. J. Math. 24 (1972), 520-529.
2. M.D. Choi, Completely positive linear maps on complex matrices. Linear Algebra and Appl. 10 (1975), 285-290.
3. M.D. Choi, M. Giesinger, J. A. Holbrook, and D.W. Kribs, Geometry of higher-rank numerical ranges, Linear and Multilinear Algebra 56 (2008), 53-64.
4. M.D. Choi, J.A. Holbrook, D. W. Kribs, and K. Życzkowski, Higher-rank numerical ranges of unitary and normal matrices, Operators and Matrices 1 (2007), 409-426.
5. M.D. Choi, D. W. Kribs, and K. Życzkowski, Higher-rank numerical ranges and compression problems, Linear Algebra Appl., 418 (2006), 828–839.
6. M.D. Choi, D. W. Kribs, and K. Życzkowski, Quantum error correcting codes from the compression formalism, Rep.. Math. Phys., 58 (2006), 77–91.
7. K.E. Gustafson and D.K.M. Rao, Numerical ranges: The field of values of linear operators and matrices, Springer, New York, 1997.
8. A. Horn, Doubly stochastic matrices and the diagonal of a rotation matrix, Amer. J. Math. 76 (1954), 620–630.
9. E. Knill and R. Laflamme, Theory of quantum error correcting codes, Phys. Rev. A 55 (1997), 900-911.
10. States, effects, and operations: fundamental notions of quantum theory, Lectures in mathematics physics at the University of Texas at Austin, Lecture Notes in Physics 190, Springer-Verlag, Berlin-Heidelberg, 1983.
11. C.K. Li and Y.T. Poon, Generalized numerical ranges and quantum error correction, to appear in J. Operator Theory., http://people.wm.edu/~cklixx/jwstarf.pdf
12. C.K. Li and Y.T. Poon, Interpolation Problems by Completely positive maps, http://arxiv.org/abs/1012.1675
13. C.K. Li, Y.T. Poon, and N.S. Sze, Condition for the higher rank numerical range to be non-empty, Linear and Multilinear Algebra 57 (2009), 365-368.

14. C.K. Li, Y.T. Poon, and N.S. Sze, Higher rank numerical ranges and low rank perturbations of quantum channels, J. Mathematical Analysis Appl. 348 (2008), 843-855.

15. C.K. Li and N.S. Sze, Canonical forms, higher rank numerical ranges, totally isotropic subspaces, and matrix equations, Proc. Amer. Math. Soc. 136 (2008), 3013-3023

16. A.W. Marshall and I. Olkin, Inequaltiies: Theo Theory of Majorizations and its Applications, Academic Press, 1979.

17. F.D. Martinez Peria, P.G. Massey, L.E. Silvestre, Weak matrix majorization, Linear Algebra Appl. 403 (2005), 343368.

18. M. Nakahara and T. Ohmi, Quantum Computing: From Linear Algebra to Physical Realizations, CRS Press, New York, 2008.

19. M.A. Nielsen and I.L. Chuang, Quantum Computation and Quantum Information, Cambridge University Press, 2000.

20. Ł. Skowronek, E. Størmer, K. Życzkowski, Cones of positive maps and their duality relations, J. Math. Phys. 50 (2009) 062106.

21. H. Woerdeman, The higher rank numerical range is convex, Linear and Multilinear Algebra 56 (2008), 65-67.

UNTYING KNOTS BY NMR:

EXPERIMENTAL IMPLEMENTATION OF AN EXPONENTIALLY FAST QUANTUM ALGORITHM FOR APPROXIMATING THE JONES POLYNOMIAL

RAIMUND MARX

Department Chemie, Technische Universität München
Lichtenbergstrasse 4, 85747 Garching, Germany E-mail: marx@tum.de

The quantum algorithm for approximating the Jones polynomial is one of the few algorithms that show an exponential speedup in comparison to classical algorithms. The lecturer will present experimental results for a small-scale approximate evaluation of the Jones polynomial by nuclear magnetic resonance (NMR). In addition, he will show how to escape from the limitations of NMR approaches that employ pseudopure states. Specifically, two spin-1/2 nuclei of natural abundance chloroform are used for those experiments. The quantum algorithm is implemented by a sequence of unitary transforms representing the trefoil knot, the figure-eight knot, and the Borromean rings. After measuring the nuclear spin state of the molecule in each case, the value of the Jones polynomial for each of the knots can be estimated.

1. Mathematical Description of NMR Spectroscopy

There are two common mathematical formalisms for the description of NMR experiments: the product operator formalism and the density operator formalism. NMR spectroscopists usually prefer the product operator formalism because of its simplicity and its descriptive analysis of what happens during the experiment. But the product operator formalism does not include effects like relaxation or "strong coupling". For NMR experiments which rely on those effects, the density operator formalism has to be used.

1.1. *From the wave function to the product operator formalism*

The following table gives an overview of different mathematical descriptions which are used for nuclear spin-1/2 systems:

spin on molecule	$$\|\psi\rangle = c_1 \|\uparrow\rangle + c_2 \|\downarrow\rangle = \begin{pmatrix} c_1 \\ c_2 \end{pmatrix}$$	wave function
spin on molecule	$$\|\psi\rangle \langle\psi\|$$	matrix representation
ensemble of "spin on molecule"	$$\rho = \overline{\|\psi\rangle \langle\psi\|}$$	average of matrix representation: density operator (ρ)
ensemble in thermal state	$$\rho \propto \begin{pmatrix} 1 & 0 \\ 0 & 1 \end{pmatrix} + \epsilon \begin{pmatrix} 1 & 0 \\ 0 & -1 \end{pmatrix}$$	density operator of thermal state
ensemble in thermal state	$$\rho = \frac{1}{2} \begin{pmatrix} 1 & 0 \\ 0 & -1 \end{pmatrix}$$	NMR relevant part of density operator
ensemble in thermal state	$$\rho = I_z$$	representation as NMR product operator

1.2. Product operator formalism

1.2.1. Description of spin states

There are 4 cartesian product operators for one-spin-1/2 NMR samples[*]:
1 (identity), I_x (x-magnetization), I_y (y-magnetization), I_z (z-magnetization)
There are 16 cartesian product operators for two-spin-1/2 NMR samples:
identity: **1**
magnetization on one spin:
I_{1x} I_{2x} I_{1y} I_{2y} I_{1z} I_{2z}
correlated magnetization on two spins:
$2I_{1x}I_{2x}$ $2I_{1x}I_{2y}$ $2I_{1x}I_{2z}$ $2I_{1y}I_{2x}$ $2I_{1y}I_{2y}$
$2I_{1y}I_{2z}$ $2I_{1z}I_{2x}$ $2I_{1z}I_{2y}$ $2I_{1z}I_{2z}$

1.2.2. Description of dynamics

Radiofrequency pulses rotate magnetization vectors:
Examples:
$$I_{1z} \xrightarrow{90^\circ_y(1)} I_{1x} \xrightarrow{90^\circ_z(1)} I_{1y}$$
$$I_{1y} \xrightarrow{90^\circ_x(1)} I_{1z}$$

---

[*]such an NMR sample consists of an ensemble of molecules in solution where each molecule contains only one spin-1/2 nucleus: e.g. an ensemble of $^{12}C^1H^{35}Cl_3$ molecules dissolved in $^{12}C^{35}Cl_4$

$$2I_{1z}I_{2z} \xrightarrow{90^\circ_y(1)} 2I_{1x}I_{2z} \xrightarrow{90^\circ_x(2)} -2I_{1x}I_{2y}$$

Weak scalar coupling acts for time $1/(2J_{12})$ between 2 spins:
$\Longrightarrow$ x- or y-magnetization rotates 90°_z
AND creates/deletes z-magnetization on other spin
$\Longrightarrow$ nothing happens, if correlated x- or y-magnetization on both spins
Examples:

$$I_{1x} \xrightarrow{1/(2J_{12})} 2I_{1y}I_{2z} \xrightarrow{1/(2J_{12})} -I_{1x}$$
$$I_{2x} \xrightarrow{1/(2J_{12})} 2I_{1z}I_{2y} \xrightarrow{1/(2J_{12})} -I_{2x}$$
$$I_{1z} \xrightarrow{1/(2J_{12})} I_{1z}$$
$$2I_{1x}I_{2y} \xrightarrow{1/(2J_{12})} 2I_{1x}I_{2y}$$

1.2.3. *Description of measurements*

Measurement (of spin #1 in a two-spin NMR system): Only the following product operators are detectable:

I_{1x} and I_{1y} (after phase correction, both operators lead to the same spectrum):

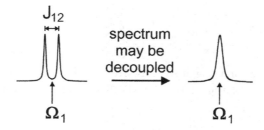

$2I_{1x}I_{2z}$ and $2I_{1y}I_{2z}$ (after phase correction, both operators lead to the same spectrum):

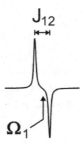

1.3. *Density operator formalism*

The density operator ρ is a matrix representation of the system. One-spin-1/2 NMR samples are represented by a 2x2 matrix. Two-spin-1/2 NMR samples are represented by a 4x4 matrix.

As most of the NMR experiments can be described with the product operator formalism as well as with the density operator formalism, there is a close connection between these two formalisms:

1.3.1. *Description of spin states*

Product operators for one-spin-1/2 NMR samples can be represented by a normalized identity matrix and normalized Pauli matrices:

$$1 = \begin{pmatrix} 1 & 0 \\ 0 & 1 \end{pmatrix}, \quad I_x = \frac{1}{2}\begin{pmatrix} 0 & 1 \\ 1 & 0 \end{pmatrix}, \quad I_y = \frac{1}{2}\begin{pmatrix} 0 & -i \\ i & 0 \end{pmatrix}, \quad I_z = \frac{1}{2}\begin{pmatrix} 1 & 0 \\ 0 & -1 \end{pmatrix}$$

Product operators for two-spin-1/2 NMR samples can be represented by the tensor product of the matrix for the first spin and the matrix of the second spin:

Examples:

$$I_{1x} = I_x \otimes 1 = \frac{1}{2}\begin{pmatrix} 0 & 1 \\ 1 & 0 \end{pmatrix} \otimes \begin{pmatrix} 1 & 0 \\ 0 & 1 \end{pmatrix} = \frac{1}{2}\begin{pmatrix} 0 & 0 & 1 & 0 \\ 0 & 0 & 0 & 1 \\ 1 & 0 & 0 & 0 \\ 0 & 1 & 0 & 0 \end{pmatrix}$$

$$2I_{1y}I_{2z} = 2 \cdot I_y \otimes I_z = 2 \cdot \frac{1}{2}\begin{pmatrix} 0 & -i \\ i & 0 \end{pmatrix} \otimes \frac{1}{2}\begin{pmatrix} 1 & 0 \\ 0 & -1 \end{pmatrix} = \frac{1}{2}\begin{pmatrix} 0 & 0 & -i & 0 \\ 0 & 0 & 0 & i \\ i & 0 & 0 & 0 \\ 0 & -i & 0 & 0 \end{pmatrix}$$

1.3.2. Description of dynamics

The effect of radiofrequency pulses is described by a unitary operator

$$U = e^{-iHt}$$

which acts on the density operator: $\rho' = U \cdot \rho \cdot U^\dagger$.
Examples:
90°_y pulse:

$$U_{90^\circ_y(1)} = e^{-i\frac{\pi}{2}I_{1y}} = \frac{1}{\sqrt{2}}\begin{pmatrix} 1 & 0 & -1 & 0 \\ 0 & 1 & 0 & -1 \\ 1 & 0 & 1 & 0 \\ 0 & 1 & 0 & 1 \end{pmatrix} \qquad U_{90^\circ_y(2)} = \frac{1}{\sqrt{2}}\begin{pmatrix} 1 & -1 & 0 & 0 \\ 1 & 1 & 0 & 0 \\ 0 & 0 & 1 & -1 \\ 0 & 0 & 1 & 1 \end{pmatrix}$$

Weak scalar coupling which acts for time $1/(2J_{12})$ between 2 spins:

$$U_J = \frac{1}{\sqrt{2}}\begin{pmatrix} 1-i & 0 & 0 & 0 \\ 0 & 1+i & 0 & 0 \\ 0 & 0 & 1+i & 0 \\ 0 & 0 & 0 & 1-i \end{pmatrix}$$

1.3.3. Description of measurements

In NMR one can only measure expectation values. The expectation value for the measurement of an operator A can be calculated as follows:
$A = \text{tr}\{A \cdot \rho\}$

1.4. For further reading

Richard R. Ernst, Geoffrey Bodenhausen, Alexander Wokaun:
"Principles of Nuclear Magnetic Resonance in One and Two Dimensions"
(International Series of Monographs on Chemistry)
Publisher: Oxford University Press, USA, 1990
ISBN-10: 0198556470; ISBN-13: 978-0198556473

2. NMR Quantum Computing Using Pseudopure States

2.1. *DiVincenzo criteria*

NMR is a method that deals with quantum mechanical systems. But can NMR be used for quantum computing? The DiVincenzo criteria are a good tool to answer that question.

2.1.1. *Qubits of an NMR-QC → (ensemble of) spin-1/2-nuclei*

The first criterion is that you should have a "scalable physical system with well characterized qubits". In NMR you deal with spin-1/2-nuclei in an external field. These two-level-systems are well characterized qubits (the issue "scalability" will be discussed later).

BUT: a typical NMR sample containes $\approx 10^{18}$ molecules which are dissolved in a liquid. And these $\approx 10^{18}$ "copies" of our qubits do NOT all do "the same". So: NMR is a method that deals with quantum mechanical systems and can therefore be used for quantum computing; but NMR quantum computing is different from "usual quantum computing" (pure state QC).

2.1.2. *Initialization of an NMR-QC → pseudopure state*

The second criterion is that you should be able to initialize the state of the qubits to a simple fiducial state, such as $|000...\rangle$.

In NMR this is just NOT possible!

BUT: we can create a pseudopure state (PPS) in our ensemble of qubits:

$\rho_{PPS} \propto \mathbb{1} + \epsilon \cdot \rho_{pure}$

The application of a non-unitary transformation can create a PPS out of the thermal state. And NMR is "blind" to identity (expectation values do not change with "amount of identity"); so in NMR a PPS "looks like" a pure state.

2.1.3. *Quantum gates: realization in an NMR-QC → sequence of r.f. pulses*

The third criterion is that you should have a "universal" set of quantum gates.

One-qubit gates (corresponding to rotation of single-spin subspace) can be implemented by a radiofrequency pulse with the correct frequency, duration, and phase.

Two qubit gates can be implemented by a radiofrequency pulse sequence which includes the evolution of weak scalar coupling. The NMR pulse sequence for a CNOT-gate between spins #1 and #2 is:

$$\xrightarrow{90^\circ_y(2)} \xrightarrow{1/(2J_{12})} \xrightarrow{90^\circ_x(2)} \xrightarrow{90^\circ_{-z}(2)} \xrightarrow{90^\circ_z(1)}$$

2.1.4. Measurement of an NMR-QC expectation value

The fourth criterion is that you should be able to measure qubit specifically. In NMR one can only measure expectation values:

Step #1: excite desired spin by 90° pulse

Step #2: acquire free induction decay (FID)

Step #3: Fourier transformation

The resulting NMR spectra (spin #1 excited) look like this (the corresponding qubit states are noted below the spectra):

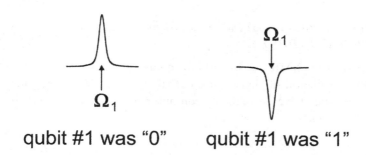

qubit #1 was "0"   **qubit #1 was "1"**

2.1.5. Coherence time of an NMR-QC → T_2 (or longer?)

The fifth criterion is that you should have long relevant decoherence times, much longer than the gate operation time.

In NMR the "decoherence time of qubits" corresponds to the T_2 relaxation time.

Typical values are:

a) For NMR systems with "tens of spins":

T_2 $(^1H) \approx 5$ s $T_2(^{13}C) \approx 1$ s $T_2(^{15}N) \approx 1$ s

b) For NMR systems with "hundreds of spins":

$T_2(^1H) \approx 50$ ms $T_2(^{13}C) \approx 500$ ms $T_2(^{15}N) \approx 500$ ms

Typical values for coupling constants and gate operation times $(\frac{1}{2J})$ are:

$^3J(^1H \text{ - } ^1H) \approx 8$ Hz → $1/(2J) \approx 63$ ms

$^1J(^{13}C \text{ - } ^{13}C) \approx 35$ Hz → $1/(2J) \approx 14$ ms

$^1J(^{15}N$ - $^{15}N) \approx 10$ Hz $\rightarrow 1/(2J) \approx 50$ ms

If you are lucky, you therefore will be able to perform:

a) For NMR systems with "tens of spins": ≈ 100 gate operations

b) For NMR systems with "hundreds of spins": ≈ 50 gate operations

There are some possible ways out of this dilemma:

- T_2 may be exceeded
  (NMR: you only expect lower signal-to-noise for $t > T_2$; not a different result)
- decrease gate operation time by using stronger couplings (residual dipolar couplings, etc.)
- increase T_2 by using supercritical solvents, etc.
- use relaxation-optimized pulse sequences

2.2. Deutsch-Jozsa quantum algorithm on a 2-qubit NMR-QC (PPS)

We just saw, that NMR can be used for quantum computing. As an example, the experimental implementation of the Deutsch-Jozsa quantum algorithm on a 2-qubit NMR quantum computer using pseudopure states will be shown explicitly:

2.2.1. Alice: initialization

a) Create PPS: $I_{1z} + I_{2z} + 2I_{1z}I_{2z}$ (corresponds to $|00\rangle$)

b) $180^\circ_y(2)$ (corresponds to flip of spin 2)

c) $90^\circ_y(1,2)$ (similar to Hadamard)

$\implies$ You then have: I_{1x} - I_{2x} - $2I_{1x}I_{2x}$

In order to create the PPS, the 2-qubit NMR-QC must be transferred from the thermal state ($I_{1z} + I_{2z}$) to the pseudopure state ($I_{1z} + I_{2z} + 2I_{1z}I_{2z}$). There are several ways to accomplish this. One of those methods is "temporal averaging":

Experiment A: use thermal state: $I_{1z} + I_{2z}$

Experiment B: create $2I_{1z}I_{2z}$ from the thermal state (and pay attention to create "the right amount" of $2I_{1z}I_{2z}$ in relation to experiment A).

The PPS experiment is the sum of both experiments.

2.2.2. *Bob: function evaluation (1-bit functions)*

A) Example for constant function:
$f_c(0) = 0$
$f_c(1) = 0$
B) Example for balanced function:
$f_b(0) = 0$
$f_b(1) = 1$

Two qubit implementation:
A) $|x\rangle |y\rangle \xrightarrow{U_{f_c}} |x\rangle |y\rangle$ "doing nothing"
$\Longrightarrow$ For the constant function, you now have: $I_{1x} - I_{2x} - 2I_{1x}I_{2x}$
B) $|x\rangle |y\rangle \xrightarrow{U_{f_b}} |x\rangle |x \oplus y\rangle$ "apply CNOT" $I_{1x} - I_{2x} - 2I_{1x}I_{2x} \xrightarrow{90°_y(2)} I_{1x}$
$+ I_{2z} + 2I_{1x}I_{2z}$
$\xrightarrow{1/(2J_{12})} 2I_{1y}I_{2z} + I_{2z} + I_{1y} \xrightarrow{90°_x(2)} -2I_{1y}I_{2y} - I_{2y} + I_{1y}$
$\xrightarrow{90°_{-z}(2)} -2I_{1y}I_{2x} - I_{2x} + I_{1y} \xrightarrow{90°_z(1)} 2I_{1x}I_{2x} - I_{2x} - I_{1x}$
$\Longrightarrow$ For the balanced function, you now have: $2I_{1x}I_{2x} - I_{2x} - I_{1x}$

2.2.3. *Alice: measurement*

a) $90°_{-y}(1,2)$ (similar to Hadamard)
b) $90°_y(1,2)$; then measurement of spin #1
$\Longrightarrow$ Pulses cancel; just measure!

$\Longrightarrow$ For the constant function, only I_{1x} is detactable on spin #1 in NMR.
$\Longrightarrow$ For the balanced function, only $-I_{1x}$ is detactable on spin #1 in NMR.
A) $\langle I_{1x}\rangle = 1$ (positive NMR signal $\to$ constant function)
B) $\langle I_{1x}\rangle = -1$ (negative NMR signal $\to$ balanced function)

2.3. *Chemical Engineering of a 5-qubit NMR quantum computer*

For 1- to 3-qubit NMR quantum computers, you can buy the compound which represents the "hardware" of the quantum computer.
Examples:
1-qubit: chloroform (one ^{1}H)
2-qubits: ^{13}C-labelled chloroform (one ^{1}H, one ^{13}C)
3-qubits: ^{13}C-labelled alanine (three ^{13}C with different frequencies)

For larger NMR-QCs one has to design and synthesize a suitable compound. It is shown in the following section, how a 5-qubit NMR-QC can be realized:

2.3.1. Design of a suitable compound ("molecule")

The following items have been considered for the design of the molecule:

- The molecule should contain five spin-1/2-nuclei.
- The coupling topology of the five spin system should be mostly linear. This is sufficient for any quantum algorithm and a more complex coupling topology would impose more difficulties for "doing nothing" to all the other qubits while manipulating only some of them.
- The couplings along this linear chain should be large - so ^1J couplings are preferable.
- The five spin-1/2-nuclei that are used as qubits should not have any couplings to other NMR active nuclei.
- As these five spin-1/2-nuclei have to be individually controllable the differences of their Larmor precession frequencies should be as large as possible - so a heteronuclear spin system is preferable.
- The compound should be easily available in sufficient amounts.
- The compound should be soluble and stable.
- The solved compound should have relaxation times (decoherence times) that are as long as possible. They should at least be longer than the time needed for the implementation of the quantum algorithm.

The compound BOC-$(^{13}C_2$-^{15}N-D_2^α-$Gly)$-F was chosen to serve as the "hardware" of the quantum computer because it fulfils most of these criteria: The molecule contains a linear chain of five spin-1/2-nuclei that are coupled via ^1J couplings and that do not couple to any other NMR active nuclei (deuterium is decoupled in all the NMR experiments). The five spin system is nearly all heteronuclear and the remaining two carbon nuclei have Larmor precession frequencies that differ by approximately 120 ppm. Starting from commercially available $^{13}C_2$-^{15}N-glycine the synthesis of the desired compound is easily performed in three steps with sufficient yields. It is soluble in deuterated DMSO and sufficiently stable to perform all the necessary experiments. The experimentally determined T_2 relaxation times for the five spins that are used as qubits are within a range from 250 ms to 590 ms, which is much longer than the 51.4 ms needed for the implemented quantum algorithm.

2.3.2. Synthesis of the chosen compound

A synthesis consisting of three steps was developed:
a) Deuteration of $^{13}C_2$-^{15}N-glycine
b) N-tert.-butoxycarbonylation of $^{13}C_2$-^{15}N- D_2^α-glycine
c) Fluorination of BOC-($^{13}C_2$-^{15}N-D_2^α)-glycine
The synthesis is shown in the following scheme:

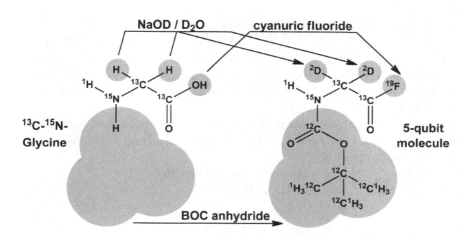

2.3.3. Coupling topology of the chosen molecule

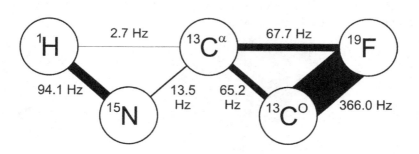

2.4. *Deutsch-Jozsa quantum algorithm on a 5-qubit NMR-QC (PPS)*

We now can use this 5-qubit compound for the experimental realization of the Deutsch-Jozsa quantum algorithm on a 5-qubit NMR quantum computer.

2.4.1. *Alice: initialization*

a) Create PPS:
The 5-qubit PPS is the sum of the following product operators:
$I_{1z}+I_{2z}+I_{3z}+I_{4z}+I_{5z}+2I_{1z}I_{2z}+2I_{1z}I_{3z}+2I_{1z}I_{4z}+2I_{1z}I_{5z}+2I_{2z}I_{3z}+2I_{2z}I_{4z}+2I_{2z}I_{5z}+2I_{3z}I_{4z}+2I_{3z}I_{5z}+2I_{4z}I_{5z}+4I_{1z}I_{2z}I_{3z}+4I_{1z}I_{2z}I_{4z}+4I_{1z}I_{2z}I_{5z}+4I_{1z}I_{3z}I_{4z}+4I_{1z}I_{3z}I_{5z}+4I_{1z}I_{4z}I_{5z}+4I_{2z}I_{3z}I_{4z}+4I_{2z}I_{3z}I_{5z}+4I_{2z}I_{4z}I_{5z}+4I_{3z}I_{4z}I_{5z}+8I_{1z}I_{2z}I_{3z}I_{4z}+8I_{1z}I_{2z}I_{3z}I_{5z}+8I_{1z}I_{2z}I_{4z}I_{5z}+8I_{1z}I_{3z}I_{4z}I_{5z}+8I_{2z}I_{3z}I_{4z}I_{5z}+16I_{1z}I_{2z}I_{3z}I_{4z}I_{5z}$
As only some of those 31 operators will contribute to the final signal, we only create: $I_{1z}+I_{2z}+I_{3z}+I_{4z}+I_{5z}+2I_{1z}I_{2z}+2I_{2z}I_{3z}+2I_{3z}I_{4z}+2I_{4z}I_{5z}$
by "temporal averaging".
b) $180_y^\circ(5)$ (corresponds to flip of spin 5)
c) $90_y^\circ(1,2,3,4,5)$ (similar to Hadamard)

2.4.2. *Bob: function evaluation (4-bit functions)*

A) Constant function: $f_0(x_1, x_2, x_3, x_4) = 0$
B) Balanced function: $f_b(x_1, x_2, x_3, x_4) = x_1 \oplus x_2 \oplus x_3 \oplus x_4$

Five qubit implementation:
$|\vec{x}\rangle |y\rangle \xrightarrow{U_f} |\vec{x}\rangle |y \oplus f(\vec{x})\rangle$
A) $|\vec{x}\rangle |y\rangle \xrightarrow{U_{f_0}} |\vec{x}\rangle |y\rangle$ "doing nothing"
B) $|\vec{x}\rangle |y\rangle \xrightarrow{U_{f_b}} |\vec{x}\rangle |x_1 \oplus x_2 \oplus x_3 \oplus x_4 \oplus y\rangle$ "apply 4 CNOTs"

NMR pulse sequence for the implementation of 4 CNOTs:

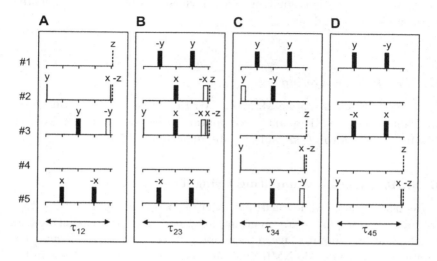

2.4.3. *Alice: measurement*

If the spectra show positive signals on spins #1 to #4; the implemented function was constant. If at least one signal on spins #1 to #4 is negative, the implemented function was balanced.

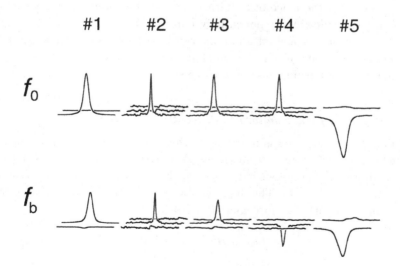

As expected we get four positive signals on spins #1 to #4 for f_0.
For f_b we get a negative signal on spin #4 which classifies the implemented function as a balanced one.

2.5. *For further reading*

R. Marx, A. F. Fahmy, John M. Myers, W. Bermel, and S. J. Glaser: "Approaching five-bit NMR quantum computing" Physical Review A, vol. 62, 012310, (2000) (http://xxx.lanl.gov/abs/quant-ph/9905087)

3. NMR Quantum Computing Using The Thermal State

In the last sections, we learned that most of the quantum algorithms that are designed for pure state quantum computing can be implemented on NMR quantum computers using pseudopure states. But even without pseudopure states one can do NMR quantum computing - as we will see in the next sections.

3.1. *Basic principles of thermal state NMR quantum computing*

This method of NMR quantum computing, which doesn't need pseudopure states, calculates the trace of a unitary matrix. Therefore, it can be used for solving the following (exponentially hard) problems:
a) Differentiation of constant and balanced functions ("Deutsch-Jozsa")
b) Evaluation of the Jones Polynomial at certain points
The quantum algorithm consists of three steps:

3.1.1. *Step 1: go from a unitary to the controlled unitary*

The aim of the algorithm is to calculate the trace of a given unitary operator which is represented by an (n by n) matrix. By adding an (n by n) identity matrix at the upper left corner of the (n by n) matrix, we created a controlled version of U. The implementation of cU needs one extra qubit as to compared with the implementation of U.

$$U \xrightarrow{\text{add one qubit}} cU = \begin{pmatrix} 1 & 0 \\ 0 & U \end{pmatrix}$$

3.1.2. *Step 2: apply cU on excited thermal state of control spin*

At first we prepare our NMR system such that we only have magnetization on our control spin (here spin #1). Then we excite the thermal state of that control spin. Finally, we apply cU.

$$cU \cdot I_{1x} \cdot cU^\dagger = \frac{1}{2} \begin{pmatrix} 0 & U^\dagger \\ U & 0 \end{pmatrix}$$

3.1.3. *Step 3: measure $\langle I_{1x} \rangle$ and $\langle I_{1y} \rangle$*

If we now measure the expectation value of I_{1x} we get twice of the real part of the desired trace of U.

$$\langle I_{1x} \rangle = \mathrm{tr}\left\{ I_{1x} \cdot \frac{1}{2} \begin{pmatrix} 0 & U^\dagger \\ U & 0 \end{pmatrix} \right\} = \frac{1}{2} \mathrm{Re}\left[\mathrm{tr}\left\{ U \right\} \right]$$

Analogous, we get twice of the imaginary part of the desired trace of U if we measure the expectation value of I_{1y}.

$$\langle I_{1y} \rangle = \mathrm{tr}\left\{ I_{1y} \cdot \frac{1}{2} \begin{pmatrix} 0 & U^\dagger \\ U & 0 \end{pmatrix} \right\} = \frac{1}{2} \mathrm{Im}\left[\mathrm{tr}\left\{ U \right\} \right]$$

3.2. *Pseudopure state vs. thermal state (pros and cons)*

The most important difference between these two methods of NMR quantum computing is the initialization: In the pseudopure state method you must perform a non-unitary transformation in order to transfer your thermal state into the pseudopure state. This transformation definitely leads to an exponential loss of signal. That is why pseudopure state NMR quantum computing is NOT scalable.

In contrast, you do not need this lossy step in thermal state NMR quantum computing. Here you directly use the thermal state and perform quantum computing with the full signal of your control spin.

The pros and cons of both methods are summarized in the following table:

	PPS	thermal state
initialization:	you loose exponentially	not necessary
scalability:	not scalable	scalable
DJ algorithm:	can be implemented	can be implemented
Grover algorithm:	can be implemented	???
Shor algorithm:	???	???
traces of matrices: e.g: Jones-Polynomial algorithm	can be implemented	can be implemented even more efficiently than on PPS or pure state QC

3.3. Deutsch-Jozsa quantum algorithm on a 2-qubit thermal state NMR-QC

As an example, the experimental implementation of the Deutsch-Jozsa quantum algorithm on a 2-qubit thermal state NMR quantum computer will be shown explicitly:

3.3.1. Alice: initialization

Alice has to prepare her NMR system such that she only has magnetization on the control spin (here spin #1). Then she has to excite the thermal state of that control spin:

a) "Deletion" of all magnetization on spins $\neq$ control spin:

$$I_{1z} + I_{2z} \xrightarrow{90_y^\circ(2)} I_{1z} + I_{2x} \xrightarrow{\text{gradient}} I_{1z}$$

b) Excitation of control spin:

$$I_{1z} \xrightarrow{90_y^\circ(1)} I_{1x}$$

3.3.2. Bob: function evaluation (1-bit functions)

Functions:

A) Example for constant function:

$f_c(0) = 0$

$f_c(1) = 0$

B) Example for balanced function:

$f_b(0) = 0$

$f_b(1) = 1$

Hamilton operators:

A) Constant function: $H_c = \begin{pmatrix} 0 & 0 \\ 0 & 0 \end{pmatrix}$

B) Balanced function: $H_b = \begin{pmatrix} 0 & 0 \\ 0 & 1 \end{pmatrix}$

Unitary operators:

A) Constant function: $U_c = \begin{pmatrix} 1 & 0 \\ 0 & 1 \end{pmatrix}$ please note: trace $\neq 0$

B) Balanced function: $U_b = \begin{pmatrix} 1 & 0 \\ 0 & -1 \end{pmatrix}$ please note: trace $= 0$

Controlled unitary operators:

A) Constant function: $cU_c = \begin{pmatrix} 1 & 0 & 0 & 0 \\ 0 & 1 & 0 & 0 \\ 0 & 0 & 1 & 0 \\ 0 & 0 & 0 & 1 \end{pmatrix}$ "doing nothing"

B) Balanced function: $cU_b = \begin{pmatrix} 1 & 0 & 0 & 0 \\ 0 & 1 & 0 & 0 \\ 0 & 0 & 1 & 0 \\ 0 & 0 & 0 & -1 \end{pmatrix}$ "controlled phase shift gate"

Now Bob has to apply the controlled unitary operator on the system, Alice has prepared:

A) Constant function: He will "do nothing"

$\implies$ The state of the NMR system will be: I_{1x}

B) Balanced function: He will apply the "controlled phase shift gate" on I_{1x}

$\implies$ The state of the NMR system will be: $\rho = cU_b \cdot I_{1x} \cdot cU_b^{\dagger}$.

$\rho = cU_b \cdot I_{1x} \cdot cU_b^{\dagger}$

$$\rho = \begin{pmatrix} 1 & 0 & 0 & 0 \\ 0 & 1 & 0 & 0 \\ 0 & 0 & 1 & 0 \\ 0 & 0 & 0 & -1 \end{pmatrix} \cdot \frac{1}{2} \begin{pmatrix} 0 & 0 & 1 & 0 \\ 0 & 0 & 0 & 1 \\ 1 & 0 & 0 & 0 \\ 0 & 1 & 0 & 0 \end{pmatrix} \cdot \begin{pmatrix} 1 & 0 & 0 & 0 \\ 0 & 1 & 0 & 0 \\ 0 & 0 & 1 & 0 \\ 0 & 0 & 0 & -1 \end{pmatrix}$$

$$\rho = \begin{pmatrix} 1 & 0 & 0 & 0 \\ 0 & 1 & 0 & 0 \\ 0 & 0 & 1 & 0 \\ 0 & 0 & 0 & -1 \end{pmatrix} \cdot \frac{1}{2} \begin{pmatrix} 0 & 0 & 1 & 0 \\ 0 & 0 & 0 & -1 \\ 1 & 0 & 0 & 0 \\ 0 & 1 & 0 & 0 \end{pmatrix}$$

$$\rho = \begin{pmatrix} 0 & 0 & 1 & 0 \\ 0 & 0 & 0 & -1 \\ 1 & 0 & 0 & 0 \\ 0 & -1 & 0 & 0 \end{pmatrix}$$

Please note, that the (2 by 2)-matrix for U_b is at the off-diagonal of ρ.

3.3.3. *Alice: measurement*

Alice now measures the expectation value of I_{1x}

A) Constant function: $\langle I_{1x} \rangle = \mathrm{tr}\,\{I_{1x} \cdot I_{1x}\} = 1$

$\Longrightarrow$ Alice will detect "full signal", which identifies the function as "constant".

B) Balanced function: $\langle I_{1x} \rangle = \mathrm{tr}\,\{I_{1x} \cdot cU_b \cdot I_{1x} \cdot cU_b^\dagger\} = \mathrm{tr}\,\{I_{1x} \cdot \rho\}$

$$\langle I_{1x} \rangle = \mathrm{tr}\left\{ \frac{1}{2}\begin{pmatrix} 0 & 0 & 1 & 0 \\ 0 & 0 & 0 & 1 \\ 1 & 0 & 0 & 0 \\ 0 & 1 & 0 & 0 \end{pmatrix} \cdot \frac{1}{2}\begin{pmatrix} 0 & 0 & 1 & 0 \\ 0 & 0 & 0 & -1 \\ 1 & 0 & 0 & 0 \\ 0 & -1 & 0 & 0 \end{pmatrix} \right\} = \mathrm{tr}\left\{ \frac{1}{4}\begin{pmatrix} 1 & 0 & 0 & 0 \\ 0 & -1 & 0 & 0 \\ 0 & 0 & 1 & 0 \\ 0 & 0 & 0 & -1 \end{pmatrix} \right\} = 0$$

$\Longrightarrow$ Alice will detect "no signal", which identifies the function as "balanced".

3.4. *Deutsch-Jozsa quantum algorithm on a 4-qubit thermal state NMR-QC*

We will use the same compound as in the 5-qubit pseudopure state NMR quantum computing example. As we only need 4-qubits this time, we will decouple spin #1 during the whole experiment. Accordingly, we will use spins # 2,3,4, and 5 for our experiment with spin # 4 as the control spin.

3.4.1. *Alice: initialization*

Alice has to prepare her NMR system such that she only has magnetization on the control spin (here spin #4). Then she has to excite the thermal state of that control spin:

a) "Deletion" of all magnetization on spins $\neq$ control spin:

$$I_{2z} + I_{3z} + I_{4z} + I_{5z} \xrightarrow{\;90^\circ_y\,(2,3,5)\;} I_{2x} + I_{3x} + I_{4z} + I_{5x}$$

$$I_{2x} + I_{3x} + I_{4z} + I_{5x} \xrightarrow{\;\text{gradient}\;} I_{4z}$$

b) Excitation of control spin:

$$I_{4z} \xrightarrow{\;90^\circ_y\,(4)\;} I_{4x}$$

3.4.2. *Bob: function evaluation (3-bit functions)*

3-bit-functions (constant / balanced):
A) Constant function:
$f_0(x_2, x_3, x_5) = 0$
B) Balanced function:
$f_b(x_2, x_3, x_5) = x_2 \oplus x_3 x_5$

4-bit-functions (after "addition" of control spin):
A) (Former) constant function:
$f'_0(x_2, x_3, x_4, x_5) = 0$
B) (Former) balanced function:
$f'_{1:3}(x_2, x_3, x_4, x_5) = (x_2 \oplus x_3 x_5) \cdot x_4 = x_2 x_4 \oplus x_3 x_4 x_5$ Controlled unitary
operators:
A) (Former) constant function:

$$U_{f'_0} = \begin{pmatrix}
1 & 0 & 0 & 0 & 0 & 0 & 0 & 0 & 0 & 0 & 0 & 0 & 0 & 0 & 0 & 0 \\
0 & 1 & 0 & 0 & 0 & 0 & 0 & 0 & 0 & 0 & 0 & 0 & 0 & 0 & 0 & 0 \\
0 & 0 & 1 & 0 & 0 & 0 & 0 & 0 & 0 & 0 & 0 & 0 & 0 & 0 & 0 & 0 \\
0 & 0 & 0 & 1 & 0 & 0 & 0 & 0 & 0 & 0 & 0 & 0 & 0 & 0 & 0 & 0 \\
0 & 0 & 0 & 0 & 1 & 0 & 0 & 0 & 0 & 0 & 0 & 0 & 0 & 0 & 0 & 0 \\
0 & 0 & 0 & 0 & 0 & 1 & 0 & 0 & 0 & 0 & 0 & 0 & 0 & 0 & 0 & 0 \\
0 & 0 & 0 & 0 & 0 & 0 & 1 & 0 & 0 & 0 & 0 & 0 & 0 & 0 & 0 & 0 \\
0 & 0 & 0 & 0 & 0 & 0 & 0 & 1 & 0 & 0 & 0 & 0 & 0 & 0 & 0 & 0 \\
0 & 0 & 0 & 0 & 0 & 0 & 0 & 0 & 1 & 0 & 0 & 0 & 0 & 0 & 0 & 0 \\
0 & 0 & 0 & 0 & 0 & 0 & 0 & 0 & 0 & 1 & 0 & 0 & 0 & 0 & 0 & 0 \\
0 & 0 & 0 & 0 & 0 & 0 & 0 & 0 & 0 & 0 & 1 & 0 & 0 & 0 & 0 & 0 \\
0 & 0 & 0 & 0 & 0 & 0 & 0 & 0 & 0 & 0 & 0 & 1 & 0 & 0 & 0 & 0 \\
0 & 0 & 0 & 0 & 0 & 0 & 0 & 0 & 0 & 0 & 0 & 0 & 1 & 0 & 0 & 0 \\
0 & 0 & 0 & 0 & 0 & 0 & 0 & 0 & 0 & 0 & 0 & 0 & 0 & 1 & 0 & 0 \\
0 & 0 & 0 & 0 & 0 & 0 & 0 & 0 & 0 & 0 & 0 & 0 & 0 & 0 & 1 & 0 \\
0 & 0 & 0 & 0 & 0 & 0 & 0 & 0 & 0 & 0 & 0 & 0 & 0 & 0 & 0 & 1
\end{pmatrix}$$

B) (Former) balanced function:

$$
U_{f'_{1:3}} =
\begin{pmatrix}
1 & 0 & 0 & 0 & 0 & 0 & 0 & 0 & 0 & 0 & 0 & 0 & 0 & 0 & 0 & 0 \\
0 & 1 & 0 & 0 & 0 & 0 & 0 & 0 & 0 & 0 & 0 & 0 & 0 & 0 & 0 & 0 \\
0 & 0 & 1 & 0 & 0 & 0 & 0 & 0 & 0 & 0 & 0 & 0 & 0 & 0 & 0 & 0 \\
0 & 0 & 0 & 1 & 0 & 0 & 0 & 0 & 0 & 0 & 0 & 0 & 0 & 0 & 0 & 0 \\
0 & 0 & 0 & 0 & 1 & 0 & 0 & 0 & 0 & 0 & 0 & 0 & 0 & 0 & 0 & 0 \\
0 & 0 & 0 & 0 & 0 & 1 & 0 & 0 & 0 & 0 & 0 & 0 & 0 & 0 & 0 & 0 \\
0 & 0 & 0 & 0 & 0 & 0 & 1 & 0 & 0 & 0 & 0 & 0 & 0 & 0 & 0 & 0 \\
0 & 0 & 0 & 0 & 0 & 0 & 0 & -1 & 0 & 0 & 0 & 0 & 0 & 0 & 0 & 0 \\
0 & 0 & 0 & 0 & 0 & 0 & 0 & 0 & 1 & 0 & 0 & 0 & 0 & 0 & 0 & 0 \\
0 & 0 & 0 & 0 & 0 & 0 & 0 & 0 & 0 & 1 & 0 & 0 & 0 & 0 & 0 & 0 \\
0 & 0 & 0 & 0 & 0 & 0 & 0 & 0 & 0 & 0 & -1 & 0 & 0 & 0 & 0 & 0 \\
0 & 0 & 0 & 0 & 0 & 0 & 0 & 0 & 0 & 0 & 0 & -1 & 0 & 0 & 0 & 0 \\
0 & 0 & 0 & 0 & 0 & 0 & 0 & 0 & 0 & 0 & 0 & 0 & 1 & 0 & 0 & 0 \\
0 & 0 & 0 & 0 & 0 & 0 & 0 & 0 & 0 & 0 & 0 & 0 & 0 & 1 & 0 & 0 \\
0 & 0 & 0 & 0 & 0 & 0 & 0 & 0 & 0 & 0 & 0 & 0 & 0 & 0 & -1 & 0 \\
0 & 0 & 0 & 0 & 0 & 0 & 0 & 0 & 0 & 0 & 0 & 0 & 0 & 0 & 0 & 1
\end{pmatrix}
$$

Now Bob has to apply the controlled unitary operator on the system, Alice has prepared:
A) (Former) constant function:
He will "do nothing"
$\Longrightarrow$ The state of the NMR system will be: I_{1x}
B) (Former) balanced function:
He will apply an NMR pulse sequence corresponding to $U(f'_{1:3})$

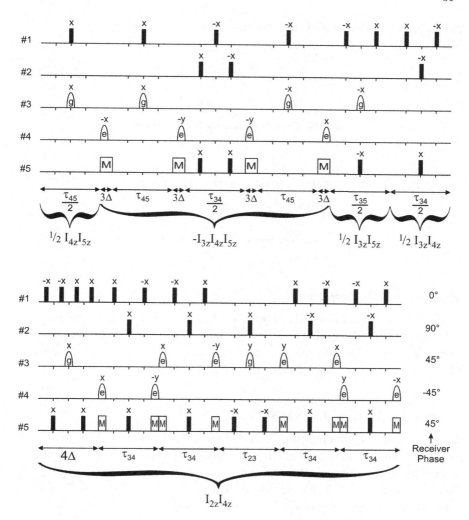

3.4.3. *Alice: measurement*

If the spectra show "full signal" when spin #4 is measured; Alice knows, that Bob implemented a controlled version of a constant function. If the spectra do not lead to any signal when spin #4 is measured; Alice knows, that Bob implemented a controlled version of a balanced function.

Measurement of $\langle I_{4x} \rangle$ leads to the following spectra:

When f_0' was implemented, we get "full signal" (as expected):

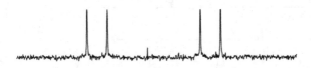

When $f_{1:3}'$ was implemented, we get "no signal" (as expected):

3.5. *For further reading*

Amr F. Fahmy, Raimund Marx, Wolfgang Bermel, and Steffen J. Glaser: "Thermal equilibrium as an initial state for quantum computation by NMR" Physical Review A, vol. 78, 022317, (2008) (http://xxx.lanl.gov/abs/0705.1676)

4. "Untying Knots by NMR"

Thermal state NMR quantum computing can be used for calculating the trace of a unitary matrix. This directly leads us to the capability of evaluating the Jones Polynomial at certain points.

In order to understand this connection between thermal state NMR quantum computing and the evaluation of Jones Polynomials, we have to consider the following details of knot theory.

4.1. *Knot theory*

4.1.1. *Definition of knots and links*

- A knot is defined as a closed, non-self-intersecting curve that is embedded in three dimensions.
- A link is defined as one or more disjointly embedded closed, non-self-intersecting curves that are embedded in three dimensions.

4.1.2. *Reidemeister moves*

One of "the big problems" in knot theory is to find out, if two given knots are equivalent? In order to clarify this, we have to define, what "equivalent" is:

Two knots / links are equivalent if they can be transformed into each other by using only Reidemeister moves and trivial moves (moves that do not change the number of crossings at any time of the move).

The following example will introduce us to the Reidemeister moves as well as a to the question, if knots are equivalent:

Question: Are these two knots equivalent?

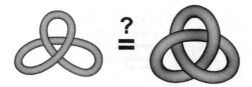

Check it: First, apply Reidemeister move #3:

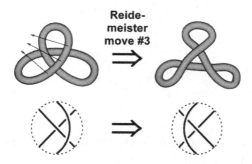

Check it: Second, apply Reidemeister move #1:

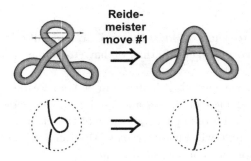

Check it: Third, apply Reidemeister move #2:

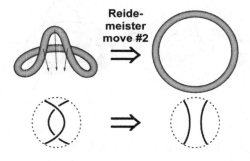

Result: No, the two given knots are NOT equivalent.

4.1.3. *Knot invariants*

Knot invariants are an important tool for the distinction if given knots are equivalent or not:

Definition:

A knot invariant is a function from the set of all knots to any other set such that the function does not change as the knot is changed (Reidemeister moves).

$\Longrightarrow$A knot invariant always assigns the same value to equivalent knots.

BUT: Different knots may have the same knot invariant.

4.1.4. *Bracket Polynomial*

The Bracket Polynomial is an important function from the set of all knots to a set of polynomials. But: is it a knot invariant?

The Bracket Polynomial is defined by:

$$\langle L \rangle (A, B, d) = \sum_{\sigma} \langle L|\sigma \rangle \, d^{||\sigma||}$$

where A and B and are the "splitting variables,"
σ runs through all "states" of the link L obtained by splitting the link,
$\langle L|\sigma \rangle$ is the "splitting label" corresponding to σ,
and $||\sigma|| \equiv N_L - 1$ where N_L is the number of loops in σ.
These rules may be illustrated as follows:

As an example, the Bracket Polynomial of the Trefoil knot is explicitely derived:

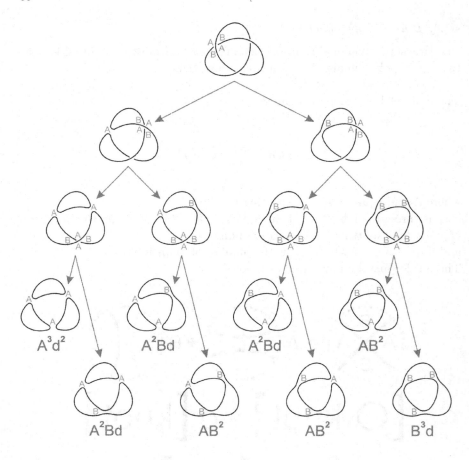

The Bracket Polynomial of the Trefoil is: $A^3d^2 + 3A^2Bd + 3AB^2 + B^3d$.

In order to find out, if the Bracket Polynomial is a knot invariant, one has to check, if it is invariant under Reidemeister moves:

- Check, if the Bracket Polynomial is invariant under Reidemeister move #2:

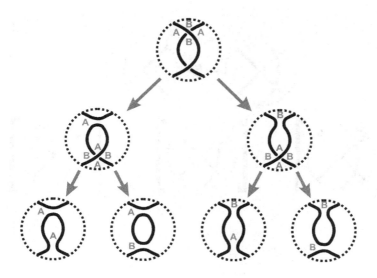

This result can be summarized as follows:

$$\left[\,\vcenter{\hbox{\includegraphics}}\,\right] = AB \left[\,\vcenter{\hbox{\includegraphics}}\,\right] + ABd \left[\,\vcenter{\hbox{\includegraphics}}\,\right] + (A^2 + B^2) \left[\,\vcenter{\hbox{\includegraphics}}\,\right]$$

Result: The Bracket Polynomial is invariant under Reidemeister move #2, if:

$AB = 1 \implies B = A^{-1}$

$ABd = -(A^2 + B^2) \implies AA^{-1}d = -(A^2 + A^{-2}) \implies d = -A^2 - A^{-2}$

- Check, if the Bracket Polynomial is invariant under Reidemeister move #3:

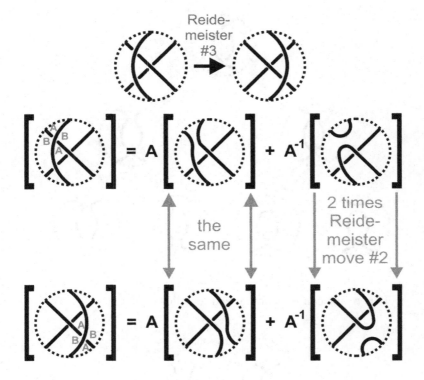

Result: The Bracket Polynomial is invariant under Reidemeister move #3.

- Check, if the Bracket Polynomial is invariant under Reidemeister move #1:

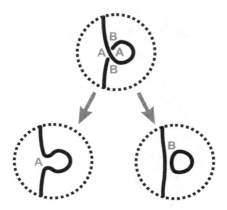

This result can be summarized as follows (remember: $B = A^{-1}$ and $d = -A^2 - A^{-2}$):

Reide-
meister
#1

$$\left[\; \circlearrowleft \;\right] = -A^{-3} \left[\;) \;\right]$$

Result: The Bracket Polynomial is NOT invariant under Reidemeister move #1. But we can make it invariant, by "adding" a term which is dependent of the writhe. This directly leads us to the Jones Polynomial.

4.1.5. *Jones Polynomial*

The Jones Polynomial is defined by:

$$X_L(A) \equiv (-A^3)^{-w(L)} \langle L \rangle (A)$$

where $\langle L \rangle (A)$ is the Bracket Polynomial depending only on one variable, where $w(L)$ is the writhe of L, defined as the sum of crossings p of L:

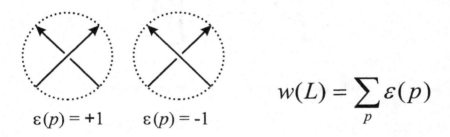

Usually, the variable is substituted as follows: $t = A^4$.
The Jones Polynomial is then expressed as: $V_L(t)$.

- The Jones Polynomial of the Unknot is equal to 1:
  $V_L(O) = 1$
- if L^* is the mirror image of L, then:
  $V_{L^*}(t) = V_L(t^{-1})$
- skein relationship (if link is oriented):
  $t^{-1} \cdot V_{L+}(t) - t \cdot V_{L-}(t) = (t^{1/2} - t^{-1/2}) \cdot V_{L0}(t)$

As an example, the Jones Polynomial of the Trefoil knot is explicitly derived: At first, the writhe of the Trefoil knot must be determined:

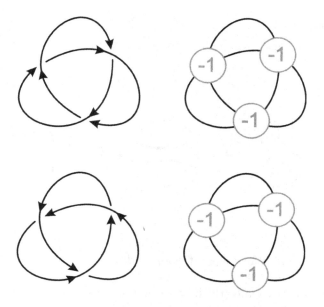

The writhe of the Trefoil is: $w = -3$.

We already determined the Bracket Polynomial of the Trefoil:
$A^3 d^2 + 3A^2 Bd + 3AB^2 + B^3 d$.

We now replace: $B = A^{-1}$ and $d = -A^2 - A^{-2}$

$\implies A^7 - A^3 - A^{-5}$

We multiply with $(-A)^{\text{negative writhe}}$

And we finally get the Jones Polynomial of the Trefoil: $-A^{16} + A^{12} + A^4$.

4.2. Different methods for the determination of the Bracket Polynomial

The key step (that is the exponentially hard step) in the determination of the Jones Polynomial ist the determination of the Bracket Polynomial. That is why it is worth looking at different methods for the determination of the Bracket Polynomial.

4.2.1. *Method #1*

"Open" all crossings of the knot according to the following rule:

with $B = A^{-1}$ and d for "separate loops": $d = -A^2 - A^{-2}$.

We already showed explicitly how to determine the Bracket Polynomial of the Trefoil using this method.

4.2.2. *Method#2*

Step #1: Describe the knot as a sequence of braid group elements which are trace-closed to build the desired knot.[†]

Step #2: "Open" the crossings of the braid group elements. The braid group representation of the knot will be transferred into a Temperley-Lieb Algebra representaion of the knot.[†]

Step #3: Perform the trace-closure (Markov trace[†]) of the sequence of Temperley-Lieb Algebra elements. The following scheme shows method #2 applied on the Trefoil:

[†]A detailed example, further definitions and explanations are shown on the following pages.

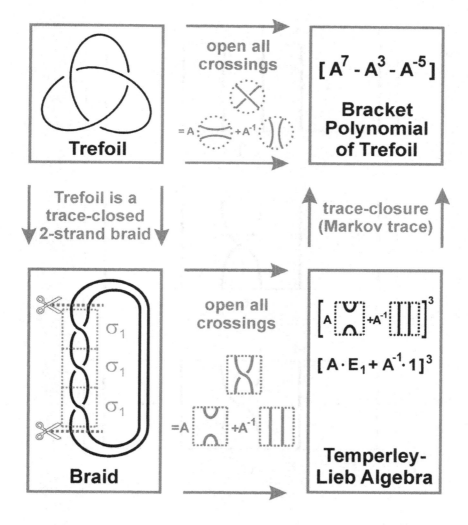

Here are some more details of how method#2 leads to the Bracket Polynomial of the Trefoil knot:

Step #1: This is the graphical representation of the generators of the n-stranded braid group B_n:

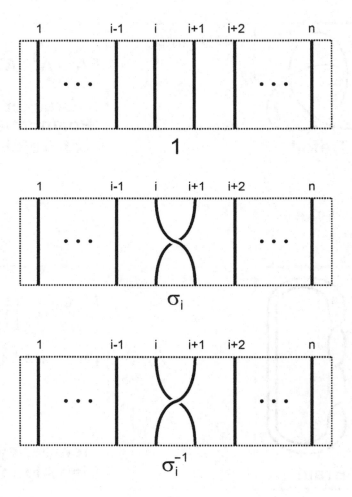

The Trefoil knot corresponds to the following sequence of B_2 braid group elements: $(\sigma_1)^3$.

Step #2: This is the graphical representation of the generators of the n-stranded Temperley-Lieb Algebra TL_n:

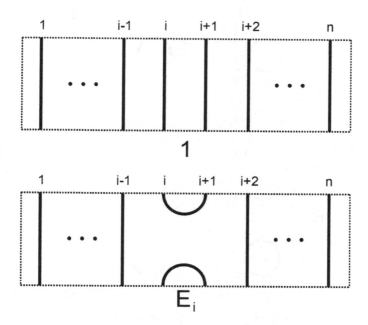

The "opening" of a σ_1 braid group element leads to the following combination of TL_2 Temperley-Lieb Algebra elements: $A \cdot E_1 + A^{-1} \cdot 1$.

As the Trefoil knot corresponds to $(\sigma_1)^3$, its sequence of Temperley-Lieb Algebra elements is:

$(A \cdot E_1 + A^{-1} \cdot 1)^3 = A^3 E_1^3 + 3A E_1^2 1 + 3A^{-1} E_1 1^2 + A^{-3} 1^3$. A graphical representation of E_1^3 and $E_1^2 1$ leads to some simplification rules:

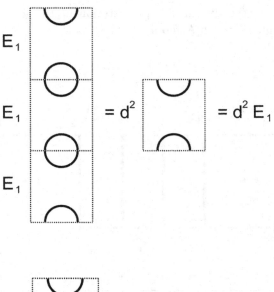

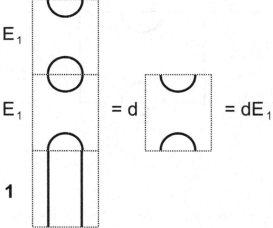

Additionally, there are the following (trivial) simplification rules:
$E_1 1^2 = E_1$ and $1^3 = 1$.

The representation of the Trefoil knot as a sequence of Temperley-Lieb Algebra elements therefore simplifies to:

$(d^2 A^3 + 3dA + 3A^{-1})E_1 + A^{-3}1$

Step #3: The Markov-trace is defined as follows:
$\operatorname{tr}\{\text{TL}_n \text{ element}\} = d^{-n} \cdot d^m$
where m is the number of "unknots" when the TL_n *element* is trace-closed.
For our example of the Trefoil knot we need to determine the Markov-trace
of both Temperley-Lieb Algebra elements in TL_2:

$$tr\left\{\;\right\} = d^{-2} \qquad = d^{-2} \cdot d^1 = d^{-1}$$

$$tr\left\{\;\right\} = d^{-2} \qquad = d^{-2} \cdot d^2 = 1$$

The correlation between the Bracket Polynomial and the Markov-trace is
defined as follows:
$\langle L \rangle(A) = d^{n-1} \cdot \operatorname{tr}\{\rho\}$
where tr is the Markov-trace and ρ is the TL_n representation of the knot
or link L.

For the Bracket Polynomial of the Trefoil knot we get:

$$
\begin{aligned}
\langle L \rangle (A) &= d^1 \cdot \mathrm{tr}\{d^2 A^3 + 3dA + 3A^{-1})E_1 + A^{-3}\mathbf{1}\} \\
&= d^1 \cdot [d^2 A^3 + 3dA + 3A^{-1})d^{-1} + A^{-3}] \\
&= A^7 - A^3 - A^{-5}
\end{aligned}
$$

4.2.3. Method#3

Steps #1 and 2 are the same as in method #2.
Step #3: Find a mapping from the Temperley-Lieb Algebra elements to unitary operators (matrices). This mapping can be accomplished by the path model representation.
Step #4: Calculate the matrix trace.

4.2.4. Method#4

Steps #1, 2, and 3 are the same as in method #3.
Step #4: Transform the unitary operators (matrices) into controlled unitary operators.
Step #5: Transform the controlled unitary operators into NMR pulse sequences.
Step #6: Apply those pulse sequences on the excited thermal state of your control spin.
Step #7: Measure the expectation value $\langle I_x \rangle$ and $\langle I_y \rangle$ of your control spin. The following scheme shows method #4 applied on the Trefoil:

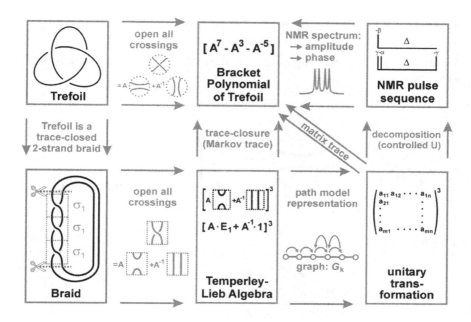

This rough scheme should become more clear, when looking at an explicit example:

4.3. *Experimental implementation of the AJL quantum algorithm on a 2-qubit thermal state NMR-QC*

In the last section, the determination of the Bracket Polynomial via NMR was shown in a very schematic way. The next section will describe an explicit example: The Bracket Polynomial of the Trefoil knot, the Figure-Eight knot and the Borromean rings (a link) will be determined using a 2-qubit thermal state NMR quantum computer.

4.3.1. *Step #1: From the knot to the sequence of braid group elements.*

a) The Trefoil knot can be represented by the following sequence of braid group elements in B_3 (three strand braid group): σ_1 - σ_1 - σ_1

b) The Figure-Eight knot can be represented by: σ_1 - σ_2^{-1} - σ_1 - σ_2^{-1}

c) The Borromean rings can be represented by:
σ_1 - σ_2^{-1} - σ_1 - σ_2^{-1} - σ_1 - σ_2^{-1}

4.3.2. *Step #2: From the sequence of braid group elements to Temperley-Lieb Algebra elements.*

We now "open" the crossing of the braid group elements according to the following rule:

with B = A⁻¹

This leads to:

$$\sigma_1 = A \cdot E_1 + A^{-1} \cdot 1$$
$$\sigma_1^{-1} = A^{-1} \cdot E_1 + A \cdot 1$$
$$\sigma_2 = A \cdot E_2 + A^{-1} \cdot 1$$
$$\sigma_2^{-1} = A^{-1} \cdot E_2 + A \cdot 1$$

4.3.3. *Step #3: From Temperley-Lieb Algebra elements to unitary operators (matrices).*

This step is too long to describe here. The method is called "path model representation". Finally, one ends up with a (2 by 2)-matrix for U_1 which corresponds to the braid group element σ_1 (so we already combined steps #2 and #3). Analogously, one can derive a (2 by 2)-matrix for U_2 which corresponds to the braid group element σ_2. These matrices contain the variable θ, which is correlated with the variable A of the Bracket polnomial.

$$U_1 = \begin{pmatrix} e^{-i\theta} & 0 \\ 0 & -e^{i\theta}\frac{\sin(4\theta)}{\sin(2\theta)} + e^{-i\theta} \end{pmatrix}$$

$$U_2 = \begin{pmatrix} -e^{i\theta}\frac{\sin(6\theta)}{\sin(4\theta)} + e^{-i\theta} & -e^{i\theta}\frac{\sqrt{\sin(6\theta)\sin(2\theta)}}{\sin(4\theta)} \\ -e^{i\theta}\frac{\sqrt{\sin(6\theta)\sin(2\theta)}}{\sin(4\theta)} & -e^{i\theta}\frac{\sin(2\theta)}{\sin(4\theta)} + e^{-i\theta} \end{pmatrix}$$

4.3.4. *Step #4: From unitary operators to controlled unitary operators.*

By adding a (2 by 2) identity matrix at the upper left corner of the (2 by 2) matrices of U_1 and U_2, we create the controlled unitary operators cU_1 and cU_2.

$$U \xrightarrow{\text{add one qubit}} cU = \begin{pmatrix} 1 & 0 \\ 0 & U \end{pmatrix}$$

4.3.5. *Step #5: From controlled unitary operators to NMR pulse sequences.*

It is well known in NMR spectroscopy, how to translate unitary operators into NMR pulse sequences (but too long to describe it here). We finally get the following pulse sequences, which also depend on the variable θ:

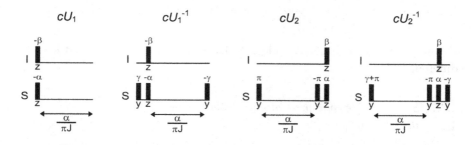

$$\text{with}: \alpha = \frac{\pi}{2} - 2\theta \qquad \beta = \frac{\pi}{2} + \theta \qquad \gamma = \arctan\left(\frac{\cos 4\theta}{\sqrt{4\cos^2 2\theta - 1}}\right) + \frac{\pi}{2}$$

4.3.6. *Step #6: Application of NMR pulse sequences on the excited thermal state of the control spin.*

We have to choose a suitable compound ("molecule") which represents the "hardware" of the 2-qubit thermal state quantum computer. A cheap and very well suitable compound is chloroform, which contains about 1% of ^{13}C chloroform. Those ^{13}C-labelled molecules will be used for quantum computing. By addition of about 10% of acetone-d_6 we will also have a sufficient amount of ^{2}D atoms in the sample, which makes the adjustment of the spectrometer easier.

The excited thermal state of the control spin was prepared by the following procedure:

a) "Deletion" of all ^{1}H-magnetization (saturation by continuous wave irradiation; 90° pulse and z-gradient; 90° pulse and y-gradient; 90° pulse and x-gradient)

b) Excitation of the ^{13}C-magnetization by a 90° pulse. Finally the desired NMR pulse sequence was applied:

- For the Trefoil knot, which can be represented as $(\sigma_1)^3$ the NMR pulse sequence for cU_1 was applied three times: $(cU_1)^3$
- For the Figure-Eight knot, which can be represented as the sequence of braid group elements $(\sigma_1 - \sigma_2^{-1} - \sigma_1 - \sigma_2^{-1})$ the corresponding NMR pulse sequence was applied: $(cU_1 - cU_2^{-1} - cU_1 - cU_2^{-1})$.
- For the Borromean Rings, which can be represented as the sequence of braid group elements $(\sigma_1 - \sigma_2^{-1} - \sigma_1 - \sigma_2^{-1} - \sigma_1 - \sigma_2^{-1})$ the corresponding NMR pulse sequence was applied: $(cU_1 - cU_2^{-1} - cU_1 - cU_2^{-1} - cU_1 - cU_2^{-1})$.

For each knot or link, the complete NMR experiment was repeated for 31 different values of θ.

4.3.7. *Step #7: Measurement of expectation values: $\langle I_x \rangle$ and $\langle I_y \rangle$ of the control spin.*

Finally, the expectation values: $\langle I_x \rangle$ and $\langle I_y \rangle$ of ^{13}C were measured. For convenience, these two values were measured in separate NMR experiments; even though it is possible to measure them in just one experiment: Here are the experimental results:

Trofoil knot:
Real (left) and imaginary (right) part of the trace of the implemented matrix.
From these results, one can calculate the Jones Polynomial:
$(A^{-4} + A^{-12} - A^{-16}) \cdot (-A^2 - A^{-2})$

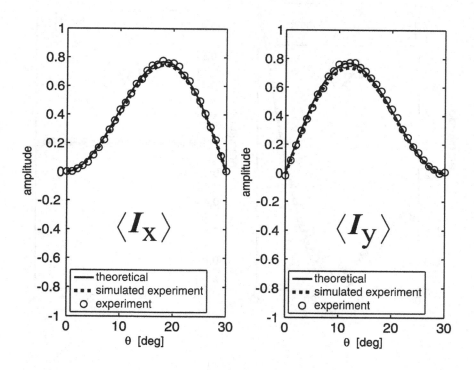

Figure-Eight knot:
Real (left) and imaginary (right) part of the trace of the implemented matrix.
From these results, one can calculate the Jones Polynomial:
$(A^8 - A^4 + A^0 - A^{-4} + A^{-8})$

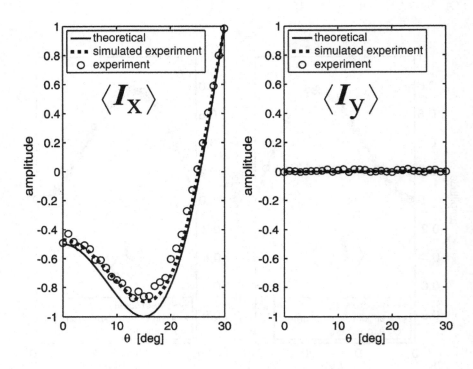

Borromean Rings:

Real (left) and imaginary (right) part of the trace of the implemented matrix.

From these results, one can calculate the Jones Polynomial:

$$(-A^3 + 3A^2 - 2A^1 + 4A^0 - 2A^{-1} + 3A^{-2} - A^{-3})$$

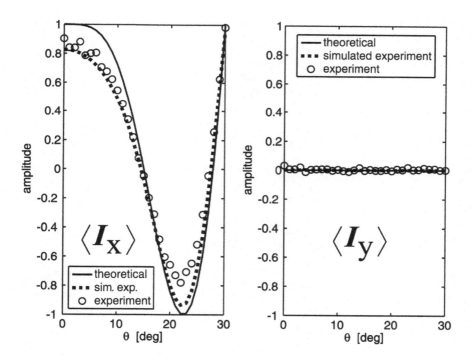

5. For Further Reading

Raimund Marx, Amr Fahmy, Louis Kauffman, Samuel Lomonaco, Andreas Spörl, Nikolas Pomplun, Thomas Schulte-Herbrüggen, John M. Myers, and Steffen J. Glaser: "Nuclear-magnetic-resonance quantum calculations of the Jones polynomial" Physical Review A, vol. 81, 032319, (2010) (http://xxx.lanl.gov/abs/0909.1080)

QUANTUM ALGORITHMS FOR PROBLEMS IN NUMBER THEORY, ALGEBRAIC GEOMETRY, AND GROUP THEORY

Wim van Dam

Department of Computer Science, Department of Physics, University of California, Santa Barbara, California, 93106-5110, USA
E-mail: vandam@cs.ucsb.edu

Yoshitaka Sasaki

Interdisciplinary Graduate School of Science and Engineering, Kinki University, Higashi-Osaka, Osaka 577-8502 Japan
E-mail: sasaki@alice.math.kindai.ac.jp

Quantum computers can execute algorithms that sometimes dramatically outperform classical computation. Undoubtedly the best-known example of this is Shor's discovery of an efficient quantum algorithm for factoring integers, whereas the same problem appears to be intractable on classical computers. Understanding what other computational problems can be solved significantly faster using quantum algorithms is one of the major challenges in the theory of quantum computation, and such algorithms motivate the formidable task of building a large-scale quantum computer. This article will review the current state of quantum algorithms, focusing on algorithms for problems with an algebraic flavor that achieve an apparent superpolynomial speedup over classical computation.

1. Introduction

Ever since Shor [20] showed in 1994 that a quantum computer can efficiently factor integers and calculate discrete logarithms, many more quantum algorithms have been discovered. In this article, we review some quantum algorithms and the problems that they try to solve such as the discrete logarithm problem, the Abelian and non-Abelian hidden subgroup problem, the point counting problem of finite field equations, and so on. A significant number of these algorithms follow the set-up of Shor's algorithm, which can be described as follows.

1. Given a function $F : G \to S$, create the superposition

$$|\Psi\rangle = \frac{1}{\sqrt{|G|}} \sum_{x \in G} |x, F(x)\rangle,$$

where G is a finite group and S is a set.

2. Apply a unitary transformation $U : \Psi \mapsto \Psi'$, such as the quantum Fourier transform.

3. Measure Ψ' in the computational basis.

Several quantum algorithms that we will discuss use notions from number theory, algebra and group theory. To help the reader's understanding of such algorithms, we will give several brief expositions of these topics. This article is based on [4] where the reader can find further details.

The article is organized as follows. In Section 2, we introduce modular arithmetic. In Sections 3 and 4 we introduce Shor's algorithm for the period finding and the discrete logarithm problem. In Section 5 we treat the Abelian hidden subgroup problem and introduce some necessary notions to understand the quantum algorithm for the Abelian hidden subgroup problem. In Section 7, we describe the elliptic curve problem and its corresponding discrete logarithm problem, and in Section 9, we discuss the quantum algorithm for efficiently solving Pellfs equation. Section 10 introduces the non-Abelian version of the quantum Fourier transform and discuss the status of the non-Abelian version of the hidden subgroup problem.

2. Modular Arithmetic and Residue Class of N

We must first introduce a little group theory for a better understanding of many quantum algorithms. A group is a combination of a set with a binary operation on its elements that obeys certain required group properties. For our purposes here, an important instance of a finite group is the residue class of integers modulo N with addition as its group operation.

Modular arithmetic, (mod N), is understood through a congruence relation on the integers $\mathbb{Z}$, where for integers a, b we have

$$a \equiv b \quad (\text{mod } N) \quad \text{if and only if } b - a \text{ ca be divided by } N.$$

Example 2.1.

- $5 \equiv 2 \equiv -1 \pmod{3}$.
- $3 \equiv 15 \pmod{12}$
- $341 \equiv 1 \pmod{10}$

Note that modular arithmetic is properly defined for addition, subtraction and multiplication. Hence we have for example, $2 \times 7 \equiv 1$ (mod 13) and $(x + y)^2 \equiv x^2 + y^2$ (mod 2).

Modulo N arithmetic classifies the integers $\mathbb{Z}$ into N different equivalence classes, which can be indicated by the residues of integers when divided by N. We thus have

$$\mathbb{Z}/N\mathbb{Z} := \{0, 1, \ldots, N - 1\}$$

where 0 represents $N\mathbb{Z} = \{\ldots, -N, 0, N, 2N, \ldots\}$, 1 represents $1 + N\mathbb{Z} = \{\ldots, -N + 1, 1, N + 1, 2N + 1, \ldots\}$, and so on.

As mentioned above, modular arithmetic satisfies additiion and multiplication. Therefore we can define such arithmetics in $\mathbb{Z}/N\mathbb{Z}$ by

$$a + b \quad (\text{mod } N), \qquad a \cdot b \quad (\text{mod } N).$$

Note that for each element a in $\mathbb{Z}/N\mathbb{Z}$, there exists an additive inverse element b (such that $a + b \equiv 0$ (mod N)) in $\mathbb{Z}/N\mathbb{Z}$, that is, $N - a$. It is straightforward to verify that addition in $\mathbb{Z}/N\mathbb{Z}$ makes up a finite Abelian group. What about multiplication in $\mathbb{Z}/N\mathbb{Z}$?

Let us compare $\mathbb{Z}/5\mathbb{Z}$ with $\mathbb{Z}/6\mathbb{Z}$. The number 2 does not have a multiplicative inverse mod6, but it does for mod5. It can be shown that a number y has a multiplicative inverse z (such that $yz \equiv 1$ (mod N)) if and only if the greatest common divisor of y and N obeys $\gcd(y, N) = 1$. If $\gcd(y, N) > 1$, then there exists a nonzero z such that $yz \equiv 0 \mod N$. For instances, $2 \times 3 \equiv 1$ (mod 5) and $2 \times 3 \equiv 0$ (mod 6). Hence, we see that the set

$$(\mathbb{Z}/N\mathbb{Z})^{\times} := \{a \in \mathbb{Z}/N\mathbb{Z} \mid \gcd(a, N) = 1\}$$

forms a group with respect to multiplication, which is called the *modulo N multiplicative group*.

3. Period Finding Implies Factoring

Perhaps the best-known application of quantum computers is its efficient solution to the problem of factoring integers. To explain a closely related algorithm for factoring integers due to Miller, we first introduce the period finding problem for the sequence $x^0, x^1, x^2, \ldots \mod N$.

For any $x \in (\mathbb{Z}/N\mathbb{Z})^{\times}$, we have an r-periodic sequence $x^0 = 1, x^1, x^2, \ldots, x^r = 1, x, \ldots$. The period r is called the *multiplicative order* of x in $\mathbb{Z}/N\mathbb{Z}$ and it is a divisor of the Euler's totient value of N

$$\phi(N) := |(\mathbb{Z}/N\mathbb{Z})^\times| = N \prod_{\substack{p|N \\ p:\text{prime}}} \left(1 - \frac{1}{p}\right),$$

which expresses the size of the multiplicative group modulo N. As an example, for $4 \in (\mathbb{Z}/9\mathbb{Z})^\times$, we have a sequence $1 = 4^0, 4 = 4^1, 7 \equiv 4^2, 1 \equiv 4^3, 4 \equiv 4^4, \ldots \pmod{9}$. Therefore, the period of 4 in $(\mathbb{Z}/9\mathbb{Z})^\times$ is 3, a divisor of $\phi(9) = 9(1 - 1/3) = 6$.

3.1. *Factoring N — Miller's Algorithm*

Theorem 3.1 (Miller, 1976). *For a given odd integer N with at least two distinct prime factors, we can determine a nontrivial factor of N as follows:*

1. *Pick up a random $a \in \{2, 3, \ldots, N-1\}$.*
2. *Compute $\gcd(a, N)$. If the result is different from 1, then it is a nontrivial factor of N, and we are done. More likely, $\gcd(a, N) = 1$, and we continue.*
3. *Using a period finding algorithm, determine the order r of a modulo N. If r is odd, the algorithm has failed, and we return to step 1. If r is even, we continue.*
4. *Compute $\gcd(a^{r/2} - 1, N)$. If the result is different from 1, then it is a nontrivial factor of N. Otherwise, return to step 1.*

By repeating this routine we can find all factors of N.

With the following lemma, Miller showed that if Step 3 can be executed efficiently, then the above algorithm as a whole is efficient.

Lemma 3.1. *Suppose a is chosen uniformly at random from $(\mathbb{Z}/N\mathbb{Z})^\times$, where N is an odd integer with at least two distinct prime factors. Then with probability at least $1/2$, the multiplicative order r of a modulo N is even, and $a^{r/2} \not\equiv -1 \pmod{N}$.*

Exercise 1 (Factoring $N = 21$). *For all $a \in \mathbb{Z}/21\mathbb{Z}$ figure out what the sequence $a^0, a^1, \ldots$ tells us about the factors of $N = 21$. Which a have $\gcd(a, 21) = 1$, What are the periods? Which a give us useful information?*

Answer. A modulo multiplication group of $\mathbb{Z}/N\mathbb{Z}$ equals $(\mathbb{Z}/21\mathbb{Z})^\times := \{a \in \mathbb{Z}/21\mathbb{Z} \mid \gcd(a, 21) = 1\} = \{1, 2, 4, 5, 8, 10, 11, 13, 16, 17, 19, 20\}$. The following table gives, for all $a \in (\mathbb{Z}/21\mathbb{Z})^\times$, the sequences of values a^j, the periods r of these sequences, and whether or not this r is useful in determining a nontrivial factor of 21:

$a \setminus j$	0	1	2	3	4	5	6	period r	useful?
1	1	1	1	1	1	1	1	1	N
2	1	2	4	8	16	11	1	6	Y
4	1	4	16	1	4	16	1	3	N
5	1	5	4	20	16	17	1	6	N
8	1	8	1	8	1	8	1	2	Y
10	1	10	16	13	4	19	1	6	Y
11	1	11	16	8	4	2	1	6	Y
13	1	13	1	13	1	13	1	2	Y
16	1	16	4	1	16	4	1	3	N
17	1	17	16	20	4	5	1	6	N
19	1	19	4	13	16	10	1	6	Y
20	1	20	1	20	1	20	1	2	N

3.2. Shor's Algorithm

Shor (1994) proved that period finding can be done efficiently with a quantum algorithm. As part of this proof, he had to show how to efficiently implement the *Quantum Fourier Transform (QFT)* over $\mathbb{Z}/N\mathbb{Z}$:

$$|x\rangle \mapsto \frac{1}{\sqrt{N}} \sum_{y \in \mathbb{Z}/N\mathbb{Z}} e^{2\pi i x y / N} |y\rangle.$$

Here is a sketch of Shor's algorithm.

Algorithm 3.1 (Period Finding). *Let* $f : \mathbb{Z}/N\mathbb{Z} \to S$ *an* r*-periodic function with* $f(x) = f(y)$ *if and only if* $(x - y)/r \in \mathbb{Z}$ *and* $r|N$.

(1) *Create the superposition*

$$\frac{1}{\sqrt{N}} \sum_{x \in \mathbb{Z}/N\mathbb{Z}} |x, f(x)\rangle.$$

(2) *By tracing out the right register, the left register is equivalent to*

$$\frac{\sqrt{r}}{\sqrt{N}} \sum_{j \in \{0, 1, \ldots, N/r - 1\}} |s + jr\rangle$$

for a random and unknown $s \in \{0, \ldots, r - 1\}$.

(3) *Apply the QFT over $\mathbb{Z}/N\mathbb{Z}$ to the quantum state, yielding*

$$\frac{1}{\sqrt{r}} \sum_{k\in\{0,1,\ldots,r-1\}} e^{2\pi i s k/r} |kN/r\rangle.$$

(4) *Sampling the state will gives us kN/r for some random and unknown $k \in \{0,\ldots,r-1\}$.*

(5) *By repeating the above procedure several times, a number of multiples of N/r will be obtained. By taking the gcd of these outcomes we learn, with high probability, the value N/r and hence r itself.*

Combining the quantum algorithm for efficiently finding periods with Miller's algorithm thus gives us an efficient algorithm for factoring integers.

4. Discrete Logarithm Problem

In Section 3 we looked at the problem of determining the period for a given a in $(\mathbb{Z}/N\mathbb{Z})^\times$ of the sequence $a^0 = 1, a, a^2, a^3, \ldots, a^{r-1}, a^r = 1, \ldots$. Here we consider the problem of finding ℓ such that $a^\ell \equiv x \mod N$ for given a and x in $(\mathbb{Z}/N\mathbb{Z})^\times$. The *discrete logarithm* of x with respect to a, denoted $\mathrm{dlog}_a(x)$, is the smallest non-negative integer ℓ such that $a^\ell \equiv x \pmod{N}$.

This dlog problem can be generalized in the following way. Let $C = \langle g \rangle = \{g^0 = 1, g, g^2, \ldots\}$ be a cyclic group generated by g. Then, the discrete logarithm base g of $x \in C$ is denoted by $\mathrm{dlog}_g(x)$ and it is again the smallest non-negative integer ℓ such that $g^\ell = x$. The *discrete logarithm problem* is the problem of calculating $\mathrm{dlog}_g x$ for a given $x \in C = \langle g \rangle$.

4.1. *Shor's Algorithm for Calculating Discrete Logarithms*

Although the problem appears to be difficult for classical computers, quantum computers can calculate discrete logarithms efficiently. We describe Shor's algorithm [20] for discrete logarithm below. For simplicity, we assume that the order of the group $N := |C|$ is known. In fact, we can determine it efficiently using Shor's algorithm for period finding over $\mathbb{Z}$ (see [4, § IV.D.]).

Algorithm 4.1 (Discrete logarithm).

(1) *Create the uniform superposition*

$$|\mathbb{Z}/N\mathbb{Z}, \mathbb{Z}/N\mathbb{Z}\rangle = \frac{1}{N} \sum_{a,b\in\mathbb{Z}/N\mathbb{Z}} |a,b\rangle.$$

(2) *Define $f : \mathbb{Z}/N\mathbb{Z} \times \mathbb{Z}/N\mathbb{Z} \to C$ as follows:*

$$f(a,b) = x^a g^b.$$

Note that $f(a,b) = f(c,d)$ if and only if $(a-c)\operatorname{dlog}_g x = (d-b)$. Compute this function in an ancilla register, giving

$$\frac{1}{N} \sum_{a,b \in \mathbb{Z}/N\mathbb{Z}} |a, b, f(a,b)\rangle.$$

(3) *Discard the ancilla register, giving the state*

$$\frac{1}{\sqrt{N}} \sum_{a \in \mathbb{Z}/N\mathbb{Z}} |a, c - a \operatorname{dlog}_g x\rangle,$$

for an unknown c.

(4) *Perform a Quantum Fourier Transform over $\mathbb{Z}/N\mathbb{Z} \times \mathbb{Z}/N\mathbb{Z}$ to the two registers and measure its state, which will yield the outcome $(\gamma \operatorname{dlog}_g x, \gamma)$ for an arbitrary $\gamma \in \mathbb{Z}/N\mathbb{Z}$.*

(5) *Repeat the above steps, thus obtaining another pair of values $(\gamma' \operatorname{dlog}_g x, \gamma')$. With probability at least $6/\pi^2 \approx 0.61$ the factors γ and γ' will be co-prime, hence with the same probability the value $\gcd(\gamma \operatorname{dlog}_g x, \gamma' \operatorname{dlog}_g x)$ will equal the desired outcome $\operatorname{dlog}_g x$.*

4.2. *Cryptographic Consequences*

Being able to calculate discrete logarithms over $\mathbb{Z}/N\mathbb{Z}$ implies being able to break the Diffie-Hellman key-exchange protocol (as well as the ElGamal protocol). Unlike our quantum algorithms, the best known classical algorithm (the Number Field Sieve) has, in both cases, a proven running time of $2^{O(\sqrt{\log N \log \log N})}$ and a conjectured running time of $2^{O(\log^{1/3} N (\log \log N)^{2/3})}$.

Note that the bounds on the classical algorithm are upper bounds, not lower bounds. To find a method of proving that there is no efficient classical algorithm for factoring or discrete logarithms, is one of the major unresolved challenges in computational number theory.

Factoring and the discrete logarithm problem are so-called *natural problems,* where the problem statement contains complete information about the problem (N or N, g, t). In contrast, *black-box problems* have part of the computational problem hidden in a black-box that must be queried to find a solution. In this setting, it was proven by Cleve [5] that the classical lower bound for the period finding problem is $\Omega(N^{1/3}/\sqrt{\log N})$, while quantum mechanically we can solve the same problem with $(\log N)^{O(1)}$ quantum queries.

5. Abelian Hidden Subgroup Problem

Algorithms 3.1 and 4.1 solve particular instances of a more general problem, the *Abelian hidden subgroup problem* (Abelian HSP). Here we will describe this problem and its efficient quantum solution in its generality.

Let G be a finite Abelian group and consider a function $F : G \to S$, where S is some finite set. We say that F *hides* the subgroup $H \leq G$ if for all $x, y \in G$ we have

$$F(x) = F(y) \text{ if and only if } x - y \in H.$$

In other words, $F(x) = F(h + x)$ holds if and only if $h \in H$. To understand this situation well, it is helpful to look at the cosets H_r of H in G, which are defined by $H_r := r + H = \{r + h \; : \; h \in H\}$. The *coset decomposition* of G in terms of H refers to a mutually disjoint set of cosets $\{H_r \; : \; r\}$ such that $\bigsqcup_r H_r = G$.

Example 5.1. Take $G = \mathbb{Z}/6\mathbb{Z} := \{0, 1, 2, 3, 4, 5\}$ and $H = \{0, 3\} \leq G$. Then we see that G can be decomposed as $G = H_0 \sqcup H_1 \sqcup H_2$, where $H_0 := 0 + H = \{0, 3\}$, $H_1 := 1 + H = \{1, 4\}$, and $H_2 := 2 + H = \{2, 5\}$ are mutually disjoint cosets of H.

Going back to the Hidden Subgroup Problem, we see that a function F that hides H is constant function on each coset $r + H$ and injective on the different cosets $r + H$. In the Abelian HSP, we are asked to find a generating set for H given the ability to query the function F. For every different kind of group G and its possible subgroups H, this is a different kind of problem.

5.1. *Character and Dual Group*

To explain the efficient quantum solution to the Abelian HSP, we have to introduce the notion of a character over the finite Abelian group. For a finite Abelian group G, a *character* over G is a function $\Psi : G \to \mathbb{C}^\times := \{z \in \mathbb{C} \mid |z| = 1\}$ with the property that for all $x, y \in G$ we have $\Psi(x + y) = \Psi(x)\Psi(y)$. The set of all possible characters over G is denoted by

$$\hat{G} := \{\Psi : G \to \mathbb{C} \mid \Psi \text{ is a character over } G\},$$

called the *dual group* of G. The *trivial character* of G is the unit function $\Psi(x) = 1$.

Example 5.2. In the case of $G = \mathbb{Z}/N\mathbb{Z}$ we have the characters $\Psi_a(x) = \mathbf{e}^{2\pi i a x / N}$ for $a \in \{0, 1, \ldots, N - 1\}$ with $a = 0$ yielding the trivial $\Psi = \mathrm{id}$.

We can also define a group operation $\circ$ on $\hat{G}$ by $\Psi_a \circ \Psi_{a+b}$, showing that $(\hat{G}, \circ)$ is isomorphic to $\mathbb{Z}/N\mathbb{Z} = G$, and hence $|G| = |\hat{G}|$.

The above isomorphism between G and $\hat{G}$ is no coincidence, as for all finite Abelian groups we have $G \simeq \hat{G}$. As another example, consider a finite cyclic group $G := \{1, g, g^2, \ldots, g^{r-1}\}$ of size r generated by g. Now the r different characters of G are the functions Ψ_a defined by $\Psi_a(g^k) = \mathbf{e}^{2\pi\imath ak/r}$ for all $a, k\{0, \ldots, r-1\}$. Again the dual group $\hat{G}$ forms a group with respect to the composite of mapping $(\Psi_a \circ \Psi_b(x) = \Psi_a(x) \cdot \Psi_b(x) = \Psi_{a+b}(x)$ for $x \in G$. As $G \simeq \mathbb{Z}/r\mathbb{Z}$ we have again $G \simeq \hat{G} \simeq \mathbb{Z}/r\mathbb{Z}$ and consequently $|\hat{G}| = |G|$.

Characters have several useful properties, the following which are fundamental tools to describe quantum algorithm for the finite Abelian HSP.

Lemma 5.1. *Let G be a finite Abelian group and H a subgroup of G. For each character Ψ on G, we have*

$$\frac{1}{|G|} \sum_{x \in G} \Psi(x) = \begin{cases} 1 & \text{if } \Psi \text{ is trivial} \\ 0 & \text{if } \Psi \text{ is nontrivial} \end{cases}$$

More specifically we have

$$\frac{1}{|H|} \sum_{x \in H} \Psi(x) = \begin{cases} 1 & \text{if } \Psi \text{ is trivial on } H \\ 0 & \text{if } \Psi \text{ is nontrivial on } H. \end{cases}$$

5.2. *Quantum Fourier Transform for Abelian Groups*

In Section 3, we have described the quantum Fourier transform over $\mathbb{Z}/N\mathbb{Z}$. This transformation can be generalized to any finite Abelian group by using the characters over G. For each finite Abelian group G, we define the unitary *quantum Fourier transform* $\mathbb{C}^{|G|} \to \mathbb{C}^{|G|}$ over G by

$$|x\rangle \mapsto \frac{1}{\sqrt{|\hat{G}|}} \sum_{\Psi \in \hat{G}} \Psi(x)|\Psi\rangle.$$

for each $x \in G$. If we know enough about G, we can implement this efficiently in $\text{poly}(\log |G|)$ steps. The quantum Fourier transform will help us solve the general Abelian hidden subgroup problem.

5.3. *Solving Abelian Hidden Subgroup Problem*

Algorithm 5.1. *Let G be a finite Abelian group and let be $F : G \to S$ be a function that hides a subgroup $H \leq G$. The following efficient quantum algorithms determines (a set of generators of) H.*

(1) *Create the superposition:*

$$\frac{1}{\sqrt{|G|}} \sum_{x \in G} |x, F(x)\rangle$$

When ignoring the right register and since F hides H, the left register can be described by

$$|s + H\rangle := \frac{1}{\sqrt{|H|}} \sum_{x \in H} |s + x\rangle$$

for an unknown s.

(2) *Apply the quantum Fourier transform over G to the left register,*

$$|s + H\rangle \mapsto \frac{1}{\sqrt{|H| \cdot |G|}} \sum_{\Psi \in \hat{G}} \left(\sum_{x \in H} \Psi(s + x) \right) |\Psi\rangle$$

$$= \sqrt{\frac{|H|}{|G|}} \sum_{\Psi \in \hat{G}} \Psi(s) \left(\frac{1}{|H|} \sum_{x \in H} \Psi(x) \right) |\Psi\rangle$$

(3) *Note that only the Ψs that are trivial on H survive the summation $x \in H$. Hence when measuring the register, we will only observe such Ψs of which there are $|G|/|H|$.*

(4) *By repeating the above procedure $(\log |G|)^{O(1)}$ times, we obtain enough information, through the observed characters that are trivial on H to reconstruct H.*

5.4. Hidden Periodicity Problem over $\mathbb{Z}$

In the previous section, we saw how the Abelian hidden subgroup problem can be solved efficiently over any known finite Abelian group. An important generalization of this problem is the *hidden periodicity problem* over $\mathbb{Z}$, where F is a function defined over $\mathbb{Z}$ with a period p, i.e. $F(x) = F(y)$ if and only if $x \equiv y \mod p$. With G the infinite Abelian group $\mathbb{Z}$ and $H = p\mathbb{Z}$ its subgroup, this is yet another instance of the Hidden Subgroup Problem, but this time for infinite groups. As explained, for example, in [4, §IV.D] Shor [20] showed how to find the period p hidden by F efficiently in time poly$(\log p)$ on a quantum computer.

5.5. Decomposing Abelian Groups

It is known that any finite Abelian group $(G, +)$ has a decomposition

$$G \simeq \mathbb{Z}/p_1^{r_1}\mathbb{Z} \oplus \cdots \oplus \mathbb{Z}/p_k^{r_k}\mathbb{Z}.$$

Suppose we are given implicitly the encoding $E : G \to S$ of a finite Abelian group $(G, +)$, with the properties:

(1) $E(g)$ is an injection (i.e. $E(h) = E(g)$ implies $g = h$).
(2) Given $E(g)$ and $E(h)$ you can compute $E(g + h)$ and $E(-g)$.

Then there exists a quantum algorithm to solve the encode E efficiently. Further, we can find the above decomposition by this algorithm. Watrous [21] generalized this quantum algorithm to all "solvable" finite groups.

6. Fields

In this section we turn out attention to quantum algorithms that deal with *fields* instead of groups. A typical example of a field is given by the set of rational numbers $\mathbb{Q}$, which is closed for addition and multiplication, and each rational number x has an additive inverse number y and a multiplicative inverse number z such that $x + y = 0$ and $x \cdot z = 1$. In general a field F is a set that is closed under an addition $(+)$ and a multiplication $(\cdot)$ operation

$$+ : F \times F \to F, \qquad (x, y) \mapsto x + y,$$
$$\cdot : F \times F \to F, \qquad (x, y) \mapsto x \cdot y$$

and that satisfies that any $a, b, c, u, v \in F$,

(1) $a + (b + c) = (a + b) + c, \quad a \cdot (b \cdot c) = (a \cdot b) \cdot c$ (Associative),
(2) $a + b = b + a, \quad a \cdot b = b \cdot a$ (Commutative),
(3) $u \cdot (a + b) = u \cdot a + u \cdot b, \quad (u + v) \cdot a = u \cdot a + v \cdot b$ (Distributive)
(4) There exist elements 0 and 1 such that $a + 0 = 0 + a = a$ and $a \cdot 1 = 1 \cdot a$ for all $a \in F$. Such elements are unique and called the zero element and the unit element, respectively.
(5) For each a in F, there exists an element b in F such that $a + b = b + a = 0$. This element is called an additional inverse element of a.
(6) For each a in $F \setminus \{0\}$, there exists an element c in F such that $a \cdot c = c \cdot a = 1$. This element is called a multiplicative inverse element of a.

Unsurprisingly, when the number of elements in F is finite, F is called a *finite field*. Given a finite size $q = |F|$ there can essentially be only one field with that size, and we denote this finite field by $\mathbb{F}_q$. It is known that when F is a finite field, it must hold that $|F| = p^n$ with p a prime integer, and $n \in \mathbb{N}$.

Example 6.1.

1. $\mathbb{Q}, \mathbb{R}, \mathbb{C}$ are fields. However $\mathbb{Z}$ is not a field as not all integers have a multiplicative inverse in $\mathbb{Z}$.
2. For any prime number p, $\mathbb{Z}/p\mathbb{Z}$ is a finite field and we often write $\mathbb{F}_p$ instead of $\mathbb{Z}/p\mathbb{Z}$.
3. For N a composite number, $\mathbb{Z}/N\mathbb{Z}$ is not a finite field.

Exercise 2. Let be $\mathbb{F}_4 = \{0, 1, x, y\}$. Write down its addition and multiplication tables.

Answer. We start with the multiplication table for all $a, b \in \mathbb{F}_4$:

$a \cdot b$	$b = 0$	$b = 1$	$b = x$	$b = y$
$a = 0$	0	0	0	0
$a = 1$	0	1	x	y
$a = x$	0	x	x^2	xy
$a = y$	0	y	yx	y^2

Our task is to determine the values on the bottom right 2×2 block in the above table. If $xy = y$, we have $x = 1$ since y has an inverse element, which implies a contradiction. Similarly xy is not equal to x and hence we have $xy = 1$. As this implies that $yx = 1$ we see that x^2 and y^2 must be y and x, respectively. Hence the following multiplication table for $a, b \in \mathbb{F}_4$ is the correct one.

$a \cdot b$	$b = 0$	$b = 1$	$b = x$	$b = y$
$a = 0$	0	0	0	0
$a = 1$	0	1	x	y
$a = x$	0	x	y	1
$a = y$	0	y	1	x

What remains is to determine the rules for addition in $\mathbb{F}_4$. As $\mathbb{F}_4$ is an extension field of $\mathbb{F}_2$ we have $1 + 1 = 2 \equiv 0 \pmod 2$, which gives the following table for $a + b$ for all $a, b \in \mathbb{F}_4$

$a + b$	$b = 0$	$b = 1$	$b = x$	$b = y$
$a = 0$	0	1	x	y
$a = 1$	1	0	$1 + x$	$1 + y$
$a = x$	x	$x + 1$	0	$x + y$
$a = y$	y	$y + 1$	$y + x$	0

As in the case of the multiplicative calculation, $1 + x$ is not equal to 0, 1 and x, and hence we have $1 + x = x + 1 = y$, which shows that $x + y = 1$

and $1 + y = y + 1 = x$. Hence we end up with this table:

$a + b$	$b = 0$	$b = 1$	$b = x$	$b = y$
$a = 0$	0	1	x	y
$a = 1$	1	0	y	x
$a = x$	x	y	0	1
$a = y$	y	x	1	0

6.1. *Field Extensions*

Note that the field of real numbers $\mathbb{R}$ is included in the field of complex numbers $\mathbb{C}$. In general, when a field K contains a field F as a subset, F is called a *subfield* of K. Conversely, K is called an *extension field* of F. In this example, the extension field K of F can be viewed as a vector space over F and its dimension is called the *degree* of K over F. The complex numbers $\mathbb{C}$ is an extension field of $\mathbb{R}$ with degree 2. Another way of understanding such algebraic extensions is by viewing $\mathbb{C}$ as $\mathbb{R}$ extended with the solution of the degree 2 equation $X^2 + 1 = 0$, which allows us to write $\mathbb{C} = \mathbb{R}(X^2 + 1)$. Similarly, the finite field $\mathbb{F}_4$ is a degree 2 extension of $\mathbb{F}_2$ with an X such that $X^2 + X = 1$. Continuing with this idea, $\mathbb{F}_4$ can be further extended to $\mathbb{F}_{16}$ and in general, $\mathbb{F}_s$ is an extension of $\mathbb{F}_q$ if and only if s is an integral power of q.

6.2. *Number Fields*

A complex number α is an *algebraic number* if α satisfies some monic polynomial with rational coefficients,

$$p(\alpha) = 0, \qquad p(x) = x^n + c_1 x^{n-1} + \cdots + c_{n-1}x + c_n = 0 \quad (c_j \in \mathbb{Q}).$$

In particular, α is an *algebraic integer* if α satisfies some monic polynomial with integral coefficients. Any (usual) integer $z \in \mathbb{Z}$ is an algebraic integer, since it is the zero of the linear monic polynomial $p(x) = x - z$.

A *number field* is an extension of $\mathbb{Q}$ with some algebraic numbers. For instance,

$$K = \mathbb{Q}(\sqrt{-5}) := \{a + b\sqrt{-5} \mid a, b \in \mathbb{Q}\}$$

is a *quadratic* number field of degree 2, since $\sqrt{-5}$ is a root of the equation $x^2 + 5 = 0$. The set of algebraic integers contained in a number field K, is called the *ring of integers* $\mathcal{O}_K$ of K. One can show that $\mathcal{O}_K$ is a ring.

For example, the ring of integers of $\mathbb{Q}$ is $\mathbb{Z}$ and that of $\mathbb{Q}(\sqrt{m})$ (m is a square-free integer) is

$$\mathcal{O}_{\mathbb{Q}(\sqrt{m})} = \{a + b\omega \mid a, b \in \mathbb{Z}\},$$

where

$$\omega = \begin{cases} \sqrt{m} & \text{if } m \equiv 2,3 \pmod 4, \\ \frac{1+\sqrt{m}}{2} & \text{if } m \equiv 1 \pmod 4. \end{cases}$$

Note that, unlike $\mathbb{Z}$, a ring of integers of K in general does not satisfy the unique factorization property. For example, $6 = 3 \times 2 = (1+\sqrt{-5})(1-\sqrt{-5})$ in $\mathcal{O}_{\mathbb{Q}(\sqrt{-5})}$.

7. Elliptic Curve Cryptography

7.1. *Elliptic Curves*

Let K be a field and consider the cubic equation $Y^2 = X^3 + aX^2 + bX + c$ with $a, b, c \in K$. If this equation is nonsingular, the corresponding *elliptic curve* $E(K)$ is the set of its solutions $(X, Y) \in K^2$ combined with "the point at infinity" $\mathcal{O}$:

$$E(K) := \{(X,Y) \in K^2 \mid Y^2 = X^3 + aX^2 + bX + c\} \cup \{\mathcal{O}\}.$$

By suitable linear transformations, any elliptic curve can be rewritten in the form of the Weierstraß equation

$$Y^2 = X^3 + \alpha X + \beta \qquad (\alpha, \beta \in K).$$

Exercise 3. Consider an elliptic curve defined by $Y^2 = X^3 + 2X + 1$ over $\mathbb{F}_5$. List the solution $(X, Y) \in \mathbb{F}_5^2$.

Answer.

$$\begin{aligned} E(\mathbb{F}_5) &:= \{(X,Y) \mid Y^2 = X^3 + 2X + 1\} \cup \{\mathcal{O}\} \\ &= \{(0,1),(0,4),(1,2),(1,3),(3,2),(3,3),\mathcal{O}\}. \end{aligned}$$

Surprisingly, for the elements of $E(K)$ we can define an addition operation. To make the definition of addition easier to understand, we will consider elliptic curves over $\mathbb{R}$ for the moment. Given two points $P, Q \in E$, their sum $P+Q$ is defined geometrically as follows. First assume that neither point is $\mathcal{O}$. Draw a line through the points P and Q or, if $P = Q$ draw the tangent to the curve at P and let denote R the third point of intersection with $E(K)$ (if the line is parallel to $X = 0$, we have the intersection $R = \mathcal{O}$). Then we define $P+Q$ by the reflection of R about the x axis, where the reflection of $\mathcal{O}$ is itself. If one of P or Q is $\mathcal{O}$, we draw a vertical line through the other point, so that $P + \mathcal{O} = P$, showing that $\mathcal{O}$ is the zero element. Reflection about the X axis corresponds to negation, so we can think of the rule as saying that the three points of intersection of a line with $E(K)$ sum to $\mathcal{O}$.

The above geometrical introduction does not necessarily make sense for other fields. Nevertheless, we can take the same structure for any field K by translating the above argument in terms of coordinates in K^2. Let be $P = (x_P, y_P)$ and $Q = (x_Q, y_Q)$. Provided $x_P \neq x_Q$, the slope of the line through P and Q is

$$\lambda = \frac{y_Q - y_P}{x_Q - x_P}$$

Computing the intersection of this line with the elliptic curve E, we find

$$x_{P+Q} = \lambda^2 - x_P - x_Q,$$
$$y_{P+Q} = \lambda(x_P - x_{P+Q}) - y_P.$$

If $x_P = x_Q$, there are two possibilities for Q: either $Q = (x_Q, y_Q) = (x_P, y_P) = P$ or $Q = (x_Q, y_Q) = (x_P, -y_P) = -P$. If $Q = -P$, then $P + Q = \mathcal{O}$. On the other hand, if $P = Q$, that is, if we are computing $2P$, then the two equalities hold with λ replaced by the slope of the tangent to the curve at P, namely, $\lambda = \frac{3x_P^2 + a}{2y_P}$, unless $y_P = 0$, in which case the slope is infinite, so $2P = \mathcal{O}$.

7.2. Elliptic Curve Cryptography

The discrete logarithm problem for an elliptic curve E defined over a finite field $\mathbb{F}$ is described as follows. Given two points P and Q in E, how many times $r \in \mathbb{N}$ do we need to add P to get

$$rP = \underbrace{P + \cdots + P}_{r} = Q \ ?$$

This problem appears harder than the discrete logarithm over $(\mathbb{Z}/N\mathbb{Z})^\times$ and the best known classical algorithm for this problem has a time complexity of $\Omega(\sqrt{|\mathbb{F}|})$. As a result, elliptic curve cryptography systems that rely on the hardness of the discrete logarithm over elliptic curves allow smaller keys. An example of such a system supported by Certicom and is used in Blackberries.

Our quantum algorithm for solving the Hidden Subgroup Problem over $(E, +)$ still applies however, and thus allows us to break this crypto-system as well. For more details on the implementation of Shor's algorithm over elliptic curves, see Proos and Zalka [18], Kaye [12], and Cheung et al. [3]

8. Counting Points of Finite Field Equations

As in the case of elliptic curve over finite fields, for $f \in \mathbb{F}_q[x, y]$ a polynomial in two variables with coefficients in $\mathbb{F}_q$, the finite set of zeros of f make a curve. Our interest lies with the number of zeros, that is, the number of points of the curve $C_f := \{(x, y) \in \mathbb{F}_q^2 | f(x, y) = 0\}$. A key parameter characterizing the complexity of this counting problem is the size q of the field $q(= p^r)$ and the genus g of the curve. For a nonsingular, projective, planar curve f, the genus is $g = (d-1)(d-2)/2$, where $d = \deg(f)$ is the degree of the polynomial. Elliptic curves have genus 1.

In the case of the classical algorithm, Schoof [19] described an algorithm to count the number of points on an elliptic curve over $\mathbb{F}_q$ in time poly$(\log q)$. Subsequent results by Pila [17], Adleman and Huang [1] generalized this result to hyper-elliptic curves, giving an algorithm with running time $(\log q)^{O(g^2 \log g)}$. For fields $\mathbb{F}_{p^r}$, Lauder and Wan [14] showed the existence of a deterministic algorithm for counting points with time complexity poly(p, r, g). All these classical algorithms are bested by the quantum algorithm that Kedlaya [13] developed, which solves the same counting problem with time complexity poly$(g, \log q)$.

8.1. *Ingredients of Kedlaya's Algorithm*

Kedlaya's algorithm is based on the relation between the *class group* of a curve C_f and the zeros of the Zeta-function of f. Let $f \in \mathbb{F}_q[x, y]$ be a polynomial and let $C_f := \{\boldsymbol{x} \in \P^2(\mathbb{F}_q) \,|\, f(\boldsymbol{x}) = 0\}$ be its smooth curve in the projective plane $\P^2(\mathbb{F}_q)$. The projective plane is defined by $\P^2(\mathbb{F}_q) := \{(x, y, z) \in \mathbb{F}_q^3 \setminus \{\mathbf{0}\}\} / \equiv$ with the equivalence $(x, y, z) \equiv (x', y', z')$ if and only if there exists an $\alpha \in \mathbb{F}_q$ such that $(x', y', z') = \alpha(x, y, z)$. For each exponent $r \in \mathbb{N}$ we define N_r to be the number of zeros of f in $\P^2(\mathbb{F}_{q^r}$:

$$N_r := |\{\boldsymbol{x} \in \P^2(\mathbb{F}_{q^r}) \mid f(\boldsymbol{x}) = 0\}|.$$

Using these N_r, the Zeta-function of a curve C is defined as the power series

$$Z_C(T) := \exp\left(\sum_{r=1}^{\infty} \frac{N_r T^r}{r}.\right)$$

It is not hard to see that knowing Z_C implies knowing the values N_r.

Kedlaya's quantum algorithm exploits the fact that the sizes of the class groups of C_f will give us information about the function $Z_C(T)$. It is known that these class groups are a finite Abelian groups and as was discussed in Section 5.5, quantum computers are able to find the structure and size of a given finite Abelian group. Through this connection from the size of the class groups, through the properties of the Zeta function, Kedlaya's algorithm determines the values N_r efficiently in terms of the $\log q$ and the degree d of the polynomial f.

9. Unit Group of Number Fields

Let be K a number field and $\mathcal{O}_K$ the ring of integers of K. The *unit group* of K is the set of all multiplicatively invertible elements of $\mathcal{O}_K$. We denote the unit group of K by

$$\mathcal{O}_K^\times := \{u \in \mathcal{O}_K \mid \exists u^{-1} \in \mathcal{O}_K \text{ such that } u \cdot u^{-1} = 1\}.$$

As an example, let us consider the case of the number field $K = \mathbb{Q}(\sqrt{5})$ and its ring of integers

$$\mathcal{O}_{\mathbb{Q}(\sqrt{5})} = \{a + b\tfrac{1+\sqrt{5}}{2} \mid a, b \in \mathbb{Z}\}.$$

The element $9 + 4\sqrt{5}$ is a unit of $\mathbb{Q}(\sqrt{5})$ as it has an inverse element in the ring of integers: $(9+4\sqrt{5})(9-4\sqrt{5}) = 1$. Furthermore, all powers $(9\pm4\sqrt{5})^k$ will be units as well. As an aside, note that these units $x \pm y\sqrt{5}$ are exactly the solutions to *Pell's equation*

$$x^2 - my^2 = 1.$$

for $m = 5$. In general it is known that for $m \in \mathbb{N}$ and the corresponding real quadratic field $K = \mathbb{Q}(\sqrt{m})$ there exists a *fundamental unit* $\varepsilon_0 \in \mathcal{O}_K^\times$ such that

$$\mathcal{O}_K^\times = \{\pm\varepsilon_0^n \mid n \in \mathbb{Z}\}$$

Quantum computers can exploit the fact that an integer $x \in \mathbb{Z}[\sqrt{m}]$ is a unit if and only if

$$x\mathbb{Z}[\sqrt{m}] = \mathbb{Z}[\sqrt{m}].$$

Hence, the function $h(z) = \mathbf{e}^z \mathbb{Z}[\sqrt{m}]$ is periodic with period $\mathcal{R} := \log \varepsilon_0$ where ε_0 is the above mentioned fundamental unit of $\mathbb{Q}(\sqrt{m})$. Hallgren [11] showed that Shor's original period finding algorithm can be extended to find this period $\mathcal{R}$ in this number field setting.

10. Non-Abelian Hidden Subgroup Problems

In the previous sections, we saw that the Abelian Fourier transform can be used to exploit the symmetry of an Abelian hidden subgroup problem, and that this essentially gave a complete solution. We would like to generalize this to non-Abelian groups.

10.1. *Hidden Subgroup Problem*

Let be G a non-Abelian group. We say that a function $F : G \to S$ *hides* a subgroup $H \leq G$ if for all $x, y \in G$

$$F(x) = F(y) \text{ if and only if } x^{-1}y \in H.$$

In other words, F is constant on left cosets $H, g_1 H, g_2 H, \ldots$ of H in G, and distinct on different left cosets. The non-Abelian hidden subgroup problem is to determine H from F. For every different kind of group G and its possible subgroups H, this is a different kind of problem.

10.2. *Example: Graph Automorphism Problem*

As an instance of the non-Abelian hidden subgroup problem, we describe the graph automorphism problem. A graph is an ordered pair $G = (V, E)$ comprising a set $V := \{1, \ldots, n\}$ of vertices or nodes together with a set E of edges, which are 2-element subsets of V such that $(i, j) \in E \subseteq V^2$ implies that i and j are connected.

Let us now consider permutations $\pi \in S_n$ of the vertices of G, which are defined by $\pi G = G' = (V, E')$ with $E' = \{(\pi(i), \pi(j)) \mid (i, j) \in E\}$. For some permutations we will have $\pi G = G' = G$, for some others we will have $G' \neq G$. An *automorphism* of a graph $G = (V, E)$ is a such a permutation with $\pi G = G$. The set of automorphisms of G is a subgroup of the symmetric group of degree S_n and we denote this automorphism group by $\text{Aut}(G)$.

The graph automorphism problem is the problem of determining all automorphisms of a given graph G. Using the function $F(\pi) = \pi G$ defined over the symmetric group, this is a Hidden Subgroup Problem over S_n.

10.3. *Some Representation Theory*

To describe potential quantum algorithms for the non-Abelian hidden subgroup problem, we have to introduce some notions of the representation theory.

Let be G a finite group. A *representation* of G over the vector space $\mathbb{C}^n$ is a function $\rho : G \to GL(\mathbb{C}^n)$ with the property $\rho(x \cdot y) = \rho(x)\rho(y)$ for any $x, y \in G$, where $GL(\mathbb{C}^n)$ is the group of all invertible, linear transformations of $\mathbb{C}^n$, called the general linear group of $\mathbb{C}^n$. We thus see that ρ is a homomorphism from the group G to the group $GL(\mathbb{C}^n)$. We can easily show that $\rho(1) = I$, which is the n-dimensional unit matrix and $\rho(x^{-1}) = \rho(x)^{-1}$. We say that $\mathbb{C}^n$ is the representation space of ρ, where

n is called its *dimension* (or *degree*), denoted d_ρ. The 1-dimensional case implies that the representations are the characters mentioned in Section 5. Note that the general linear group $GL(\mathbb{C}^n)$ is also non-abelian for $n \geq 2$. For all finite groups, the representation ρ is a unitary representation, that is, one for which $\rho(x)^{-1} = \rho(x)^\dagger$ for all $x \in G$.

Given two representations $\rho : G \to V$ and $\rho' : G \to V'$, we can define their direct sum, a representation $\rho \oplus \rho' : G \to V \oplus V'$ of dimension $d_{\rho \oplus \rho'} = d_\rho + d_{\rho'}$. The representation matrices of $\rho \oplus \rho'$ are of the form

$$(\rho \oplus \rho')(x) = \begin{pmatrix} \rho(x) & 0 \\ 0 & \rho'(x) \end{pmatrix}$$

for all x in G. A representation is *irreducible* if it cannot be decomposed as the direct sum of two other representations. Any representation of a finite group G can be written as a direct sum of irreducible representations of G. We denote a complete set of irreducible representations of G by $\hat{G}$.

Another way to combine two representations is the tensor product. The tensor product of $\rho : G \to V$ and $\rho' : G \to V'$ is $\rho \otimes \rho' : G \to V \otimes V'$, a representation of dimension $d_{\rho \otimes \rho'} = d_\rho d_{\rho'}$.

Exercise 4. Consider the non-Abelian group $G = \langle A, B \rangle$ generated, under multiplication, by the following two matrices A and B:

$$A = \begin{pmatrix} 0 & 1 & 0 \\ 0 & 0 & 1 \\ 1 & 0 & 0 \end{pmatrix}, \quad B = \begin{pmatrix} 1 & 0 & 0 \\ 0 & 0 & 1 \\ 0 & 1 & 0 \end{pmatrix}.$$

How many elements does this group have? A representation is called *faithful* if for all $x \neq y$ we have $\rho(x) \neq \rho(y)$, i.e. when ρ is injective. Find a faithful representation with $d = 2$ for $G\langle A, B \rangle$.

Answer. Observe the left-action and right-action of A:

$$AB = \begin{pmatrix} 0 & 0 & 1 \\ 0 & 1 & 0 \\ 1 & 0 & 0 \end{pmatrix}, \quad BA = \begin{pmatrix} 0 & 1 & 0 \\ 1 & 0 & 0 \\ 0 & 0 & 1 \end{pmatrix}.$$

Hence, $ABA = B$. Note that $A^3 = I_3$ and $B^2 = I_3$. Therefore, any element of $\langle A, B \rangle$ can be expressed as $A^l B^k$ with $l \in \{0, 1, 2\}$ and $k \in \{0, 1\}$, which means that $|\langle A, B \rangle| = 6$.

The second problem is how to find a representation ρ over $\langle A, B \rangle$. For this purpose, we exploit the algebraic properties $A^3 = I_3$ and $B^2 = I_3$. Since ρ satisfies $\rho(XY) = \rho(X)\rho(Y)$ and $\rho(I_3) = I_2$, the images of A and B for ρ must satisfy $\rho(A)^3 = I_2$ and $\rho(B)^2 = I_2$. Therefore, this problem is equivalent to finding elements $\rho(A) = S$ and $\rho(B) = T$ in $GL_2(\mathbb{C})$, such that $S^3 = I_2$, $T^2 = I_2$ and all $S^l T^k$ are different. These requirements are met, for example, by the following two matrices

$$S = \begin{pmatrix} e^{2\pi i/3} & 0 \\ 0 & e^{-2\pi i/3} \end{pmatrix} \quad \text{and} \quad T = \begin{pmatrix} 0 & 1 \\ 1 & 0 \end{pmatrix}.$$

10.4. Non-Abelian Fourier Transform

For a finite non-Abelian group G, we define the $|G|$ dimensional quantum Fourier transform over G for every $x \in G$ by

$$|x\rangle \mapsto \frac{1}{\sqrt{|G|}} \sum_{\rho \in \hat{G}} d_\rho |\rho, \rho(x)\rangle,$$

where $|\rho\rangle$ is a state that labels the irreducible representations, and $|\rho(x)\rangle$ is a normalized d_ρ^2-dimensional state whose amplitudes are given by the matrix entries of the $d_\rho \times d_\rho$ matrix $\rho(x)$:

$$|\rho(x)\rangle := (\rho(x) \otimes I_{d_\rho}) \sum_{j=1}^{d_\rho} \frac{|j, j\rangle}{\sqrt{d_\rho}} = \sum_{j,k=1}^{d_\rho} \frac{\rho(x)_{j,k}|j, k\rangle}{\sqrt{d_\rho}}.$$

It can be shown that this quantum Fourier transform over G is a unitary matrix

$$\sum_{x \in G} |\hat{x}\rangle\langle x| = \sum_{x \in G} \sum_{\rho \in \hat{G}} \sqrt{\frac{d_\rho}{|G|}} \sum_{j,k=1}^{d_\rho} \rho(x)_{j,k} |\rho, j, k\rangle\langle x|.$$

Note that the Fourier transform over a non-Abelian G is not uniquely defined, rather, it depends on a choice of basis for each irreducible representation of dimension greater than 1.

10.5. Fourier Sampling

Applying the Fourier transformation to a superposition, we obtain

$$\sum_{x \in G} \alpha_x |x\rangle \mapsto \sum_{\rho \in \hat{G}} \sqrt{\frac{d_\rho}{|G|}} \sum_{j,k=1}^{d_\rho} \left(\sum_{x \in G} \alpha_x \rho(x)_{j,k} \right) |\rho, j, k\rangle.$$

For Abelian groups all dimensions d_ρ are 1, hence in that setting we focus only on the ρ. When generalizing this approach, where the (j, k) registers are ignored, to the non-Abelian case we speak of *weak Fourier sampling*. If the hidden subgroup H is *normal* (i.e. if for al $x \in G$ we have $xH = Hx$), then weak Fourier sampling will solve our HS Problem.

However, in the majority of non-Abelian hidden subgroup problems, weak Fourier sampling does not provide sufficient information to recover the hidden subgroup. For example, weak Fourier sampling fails to solve the HSP in the symmetric group (Grigni et al. [8], Hallgren et al. [10]) and the dihedral group. To obtain more information about the hidden subgroup, we have to focus on not only the ρ register but also the j and k registers. Such an approach is dubbed as *strong Fourier sampling*; see [4, § VII.C.,§ VII.D]. For some groups, it turns out that strong Fourier sampling of single registers simply fails. Moore, Russell and Schulman [16] showed that, regardless of what basis is chosen, strong Fourier sampling provides insufficient information to solve the HSP in the symmetric group if you restrict yourself to measurements on single measurements of (ρ, j, k) registers.

10.6. *Example: Dihedral/Hidden Shift Problem*

The hidden shift problem (also known as the hidden translation problem) is a natural variant of the hidden subgroup problem. In the hidden shift problem, we are given two injective functions $f_0 : G \to S$ and $f_1 : G \to S$, with the promise that

$$f_0(g) = f_1(sg) \quad \text{for some } s \in G.$$

The goal of the problem is to find s, the *hidden shift*.

Consider the case of the dihedral group D_n, which is the group of symmetries of a n-sides regular polygon, including both rotations and reflections generated by the following two matrices

$$A = \begin{pmatrix} \cos(2\pi/n) & \sin(2\pi/n) \\ \sin(2\pi/n) & \cos(2\pi/n) \end{pmatrix} \qquad : \text{rotation,}$$

$$B = \begin{pmatrix} 1 & 0 \\ 0 & -1 \end{pmatrix} \qquad : \text{reflection.}$$

It is known that the dihedral group D_n is equivalent to a semidirect product of $\mathbb{Z}/n\mathbb{Z}$ and $\mathbb{Z}/2\mathbb{Z}$ denoted by $\mathbb{Z}/n\mathbb{Z} \rtimes \mathbb{Z}/2\mathbb{Z}$. Roughly speaking, $\mathbb{Z}/n\mathbb{Z} \rtimes \mathbb{Z}/2\mathbb{Z}$ is the set of a direct product of $\mathbb{Z}/n\mathbb{Z}$ and $\mathbb{Z}/2\mathbb{Z}$ whose product is defined as

$$(l_1, k_1) \cdot (l_2, k_2) = (l_1 + (-1)^{k_1} l_2, k_1 + k_2)$$

for any (l_1, k_1) and (l_2, k_2) in $\mathbb{Z}/n\mathbb{Z} \times \mathbb{Z}/2\mathbb{Z}$.

We define a function F on $\mathbb{Z}/n\mathbb{Z} \rtimes \mathbb{Z}/2\mathbb{Z}$ with the property

$$F(x, 0) = F(x + s, 1) \quad \text{for some } s \in \mathbb{Z}/N\mathbb{Z}.$$

Hence, $F(\mathbb{Z}/n\mathbb{Z}, 1)$ is an s-shifted version of $F(\mathbb{Z}/n\mathbb{Z}, 0)$. Ettinger and Høyer [7] showed that this hidden shift problem can be solved with only $(\log n)^{O(1)}$ quantum queries to the function F, but it remains an open problem whether this solution can be achieved in a manner that also efficient in its time complexity.

10.7. *Pretty Good Measurement Approach to HSP*

The idea of the *pretty good measurement* is borrowed from quantum optics, and it gives a general approach to the problem of distinguishing quantum states from each other. In a pretty good measurement, for a given set of possible mixed states $\{\rho_1, \ldots, \rho_M\}$, we use the measurement operators $\Pi_1, \ldots, \Pi_M$ defined by

$$\Pi_i := \frac{1}{\sqrt{\sum_{j=1}^{M} \rho_j}} \cdot \rho_i \cdot \frac{1}{\sqrt{\sum_{j=1}^{M} \rho_j}}.$$

For many hidden symmetry problems, the PGM gives the measurement that extracts the hidden information in the most efficient way possible from the states. It also defines a specific measurement that one can try to implement efficiently.

11. Paths Towards Finding New Quantum Algorithms

Finding new quantum algorithms has proven to be a hard problem. For students and other researchers brave enough to nevertheless try to expand our current set of efficient algorithms, the following three approaches are suggested.

Find more applications of the Abelian HSP

The efficient quantum solution to the Abelian HSP should have more applications than we currently are aware of. By learning more about number theory, commutative algebra, and algebraic geometric it should be possible to discover computational problems in those fields that can also be solved efficiently in the Abelian HSP framework.

Find more Non-Abelian Groups

The non-Abelian HSP for the symmetric group and the dihedral group have well-known connections to problems in graph theory and the theory of lattices. Unfortunately we do not know, at the moment, how to efficiently solve the HSP for these groups. Find computational problems that depend on the HSP for groups that we *do* know how to solve efficiently quantum mechanically.

Find more Unitary matrices

Step away from the HSP framework and it Fourier transform alltogether and look at other unitary transformations and see what computational problems are a match for these quantum transformations.

Appendix A. Group Theory

Groups are an important notion in algebra and they are defined as follows.

Definition A.1 (Group). *A group is a set G together with a binary operation $\circ$ on G such that the following three axioms hold:*

(1) $(a \circ b) \circ c = a \circ (b \circ c)$ *(association) holds for any a, b, c in G.*
(2) *There exists an element e which satisfies $a \circ e = e \circ a = a$ for any a in G. Such element is unique and called the identity element.*
(3) *For each a in G, there exists an element b in G such that $a \circ b = b \circ a = e$, where e is the identity element. Such an element is called an inverse element of a, denoted a^{-1}.*

Remark A.1. A group G is *Abelian* (or *commutative*), if its group operation commutes (i.e. for all $a, b \in G$ we have $a \circ b = b \circ a$).

Example A.1. The set of all natural numbers $\mathbb{N}$ does not form a group with respect to the addition operation, because not all elements $x \in \mathbb{N}$ have an inverse $-x$ in $\mathbb{N}$.

Example A.2. The set of all integers $\mathbb{Z}$ forms a group with respect to addition, since the identity element is 0 and each element a has an inverse element $-a$. We can easily see that the addition on $\mathbb{Z}$ is associative. However $\mathbb{Z}$ is not a group with respect to multiplication, as its multiplicative inverses are not elements of $\mathbb{Z}$.

Example A.3. Let be

$$M_2(\mathbb{C}) := \left\{ A = \begin{pmatrix} a & b \\ c & d \end{pmatrix} \Big| a, b, c, d \in \mathbb{C} \right\},$$

$$GL_2(\mathbb{C}) := \{ A \in M_2(\mathbb{C}) \mid \det A \neq 0 \}.$$

Note that $GL_2(\mathbb{C})$ is a subset of $M_2(\mathbb{C})$. We see that $GL_2(\mathbb{C})$ forms a group with respect to the matrix multiplication, although $M_2(\mathbb{C})$ does not form a group under that operation. The matrix multiplication is associative, and there exists the unit matrix I_2 that plays the role as the identity element. In $M_2(\mathbb{C})$ however, some elements $A \in M_2(\mathbb{C})$ do not have an inverse A^{-1}.

References

1. L. M. Adleman, and M.-D. Huang, *Counting points on curves and Abelian varieties over finite fields*, Journal of Symbolic Computation, Volume 32 (2001), 171–189. preliminary version in ANTS-II 1996.
2. M. Boyer, G. Brassard, P. Høyer and A. Tapp, *Tight bounds on quantum searching*, Fortschritte der Physik, Volume 46 (1998), 493–505.
3. D. Cheung, D. Maslov, J. Mathew and D. Pradhan, *On the design and optimization of a quantum polynomial-time attack on elliptic curve cryptography*, Proceedings of the 3rd Workshop on Theory of Quantum Computation, Communication, and Cryptography, volume 5106 of Lecture Notes in Computer Science, (2008) pp. 96–104.
4. A. Childs and W. van Dam, *Quantum algorithms for algebraic problems*, Reviews of Modern Physics, Volume 82 (2010) 1–52.
5. R. Cleve, *The query complexity of order-finding*, Inf. Comput., Volume 192 (2004), 162–171, preliminary version in CCC 2000, eprint quant-ph/9911124.
6. R. Crandall and C. Pomerance, *Prime Numbers: A computational perspective*, Springer-Verlag, Berlin, 2005.
7. M. Ettinger and P. Høyer, *On quantum algorithms for noncommutative hidden subgroups*, Advances in Applied Mathematics, Volume 25 (2000), pp. 239–251.
8. M. Grigni, L.J. Schulman, M. Vazirani and U. Vazirani, *Quantum mechanical algorithms for the nonabelian hidden subgroup problem*, Combinatorica, Volume 24 (2004), 137–154, preliminary version in STOC 2001.
9. L. Grover, *A fast quantum-mechanical algorithm for database search*, Proceedings of the 28th Annual ACM Symposium on Theory of Computing (STOC '96), 1996, pp. 212–219.
10. S. Hallgren, A. Russell, and A. Ta-Shma, *The hidden subgroup problem and quantum computation using group representations*, SIAM Journal on Computing, Volume 32 (2003), 916–934, preliminary version in STOC 2000.
11. S. Hallgren, *Polynomial-time quantum algorithms for Pell's equation and the principal ideal problem*, Journal of the ACM, Volume 54 (2007), preliminary version in STOC 2002.

12. P. Kaye, *Optimized quantum implementation of elliptic curve arithmetic over binary fields*, Quantum Information and Computation, Volume 5 (2005), 474–491.

13. K. S. Kedlaya, *Quantum computation of zeta functions of curves*, Computational Complexity, Volume 15 (2006), 1–19.

14. A. Lauder and D. Wan, *Counting points on varieties over finite fields of small characteristic*, Algorithmic Number Theory, ed. J. Buhler and P. Stevenhagen, Cambridge University Press, volume 44 of Mathematical Sciences Research Institute Publications, 2002.

15. R. Lidl and H. Niederreiter, *Finite Fields*, Encyclopedia of Mathematics and Its Applications, Vol. 20, Cambridge Univ. Press, Cambridge, 1997.

16. C. Moore, A. Russell and L. J. Schulman, *The symmetric group defies strong Fourier sampling*, Proceedings of the 46th IEEE Symposium on Foundations of Computer Science, 2005, pp. 479–490.

17. J. Pila, *Frobenius maps of abelian varieties and finding roots of unity in finite fields*, Mathematics of Computation, Volume 55 (1990), 745–763.

18. J. Proos and C. Zalka, *Shor's discrete logarithm quantum algorithm for elliptic curves*, Quantum Information and Computation, Volume 3 (2003), 317–344.

19. R. Schoof, *Elliptic curves over finite fields and the computation of square roots mod p*, Mathematics of Computation, Volume 44 (1985), 483–494.

20. P. Shor, *Polynomial-time algorithms for prime factorization and discrete logarithms on a quantum computer*, SIAM Journal on Computing, Volume 26 (1997), 1484–1509.

21. J. Watrous, *Quantum algorithms for solvable groups*, Proceedings of the 33rd ACM Symposium on Theory of Computing, 2001, pp. 60–67.

LECTURE SERIES ON RELATIVISTIC QUANTUM INFORMATION

IVETTE FUENTES

*School of Mathematical Sciences, University of Nottingham,
Nottingham NG7 2RD United Kingdom*

The insight that the world is fundamentally quantum mechanical inspired the development of quantum information theory. However, the world is not only quantum but also relativistic, and indeed many implementations of quantum information tasks involve truly relativistic systems. In this lecture series I consider relativistic effects on entanglement in flat and curved spacetimes. I will emphasize the qualitative differences to a non-relativistic treatment, and demonstrate that a thorough understanding of quantum information theory requires taking relativity into account. The exploitation of such relativistic effects will likely play an increasing role in the future development of quantum information theory. The relevance of these results extends beyond pure quantum information theory, and applications to foundational questions in cosmology and black hole physics will be presented.

1. Introduction

The main purpose of the research in the field of relativistic quantum information is developing quantum information theory that is compatible with the relativistic structure of spacetime. An ultimate aim is to exploit relativity in order to improve quantum information tasks. The vantage point of these investigations is that the world is fundamentally both quantum and relativistic. Impressive technological achievements and promises have already been derived from taking seriously solely the quantum aspects of matter: quantum cryptography and communication have become a technical reality in recent years, but the practical construction of a quantum computer still requires to understand how to efficiently store, manipulate and read information, without prohibitively large disturbances from the environment. Throwing relativity into the equation fundamentally changes the entire game, as I intend to show in this series of lectures. Hopefully, we will be able to push this exciting line of theoretical research to the point where relativistic effects in quantum information theory can be exploited technologically.

Far from yielding only quantitative corrections, relativity plays a dominant role in the qualitative behavior of many physical systems used to implement quantum information tasks in the laboratory. The prototypical example is provided by any system involving photons, be it for the transmission or manipulation of quantum information. There is no such thing as a non-relativistic approximation to photons, since these always travel at the speed of light. While relativistic quantum theory, commonly known as quantum field theory, is a very well studied subject in foundational particle physics, research in quantum information theory selectively focused almost exclusively on those aspects one can study without relativity. Thus both unexpected obstacles (such as relativistic degradation of entanglement) and unimagined possibilities for quantum information theory (such as improved quantum cryptography and hypersensitive quantum measurement devices) have gone unnoticed. Moreover, the impact of the work done in the field of relativistic quantum information extends beyond pure quantum information theory, and applications to foundational questions in cosmology and black hole physics have been found.

1.1. *Background*

Quantum information was first considered in a relativistic setting by Czachor in 1997 who analyzed a relativistic version of the Einstein-Podolsky-Rosen-Bohm experiment.[2,3] Czachor's work showed that relativistic effects are relevant to the experiment where the degree of violation of Bell's inequalities depends on the velocity of the entangled particles. In 2002 A. Peres & D. Terno at Technion pointed out that most concepts in quantum information theory may require a reassessment.[4] Further interesting results on entanglement in flat spacetime were obtained by Adami, Bergou, & Gingrich at Caltech, and Solano & Pachos at the Max-Plank Institute.[5] Their work shows that although entanglement is overall conserved under a change of inertial frame, it may swap between spin and position degrees of freedom. For the physically interesting case of non-inertial frames, however, collaborators and I were able to show in a series of papers[6-8] that entanglement is observer-dependent, since it is degraded from the perspective of observers in uniform acceleration. This effect introduces errors in entanglement-based quantum information tasks in non-inertial frames such as the teleportation scheme studied by P. Alsing (University of New Mexico) & G. Milburn (University of Queensland).[9,10] Errors induced by relativistic effects were also found in a cryptographic protocol analyzed by M. Czachor and M. Wilczewski at Politechnika Gdańska.[11] However, J. Barrett (Universite Libre de Bruxelles), L. Hardy (Perimeter Institute), and A. Kent (University of Cambridge) showed that constraints imposed by relativity can be useful in ensuring security in quantum cryptography even if quantum theory is incorrect.[12] Considering quantum information in curved spacetime is more complicated since generically particle states are ill-defined if the background spacetime does not at least feature a timelike Killing vector field. In spacetimes with two asymptotically flat regions, however, we showed that measurements of entanglement may be used to learn about the history of spacetime.[13] This work was followed up by G. Ver Steeg (Caltech) & N. Menicucci (Princeton).[14] Entanglement in spacetime has also been studied by Shi at Tsinghua University[15] and by P. Kok, U. Yurtsever, S. L. Braunstein, and J. P. Dowling at Caltech and Bangor University.[16]

Employing quantum information to address open questions in other fields, gravity being a particularly fruitful example, has become important to quantum information scientists. J. Preskill (Caltech), D. Gottesman (Perimeter Institute), D. Ahn (University of Soul), S. Lloyd (MIT), G. Adesso and myself, among others, have shown that quantum information is a useful tool in the understanding of the information loss problem in black holes.[17] Quantum information has also been recently employed by D. Terno (Macquarie University), E. Livine (ENS-Lyon) and F. Markopoulou (Perimeter Institute) to make progress in quantum gravity.[18] Relativistic quantum information starts to be a topic of interest in the scientific community. However, technological applications of relativistic quantum information are yet to be proposed.

2. Motivation and Technical Tools in Quantum Information

2.1. *Why relativistic qunatum information?*

In information theory we study how to send messages, how to transmit information in secure ways, how to compute and in general, how to process information. In order to process information, information must be stored in physical systems. Therefore, it is the underlying physical theory which sets the rules for which information tasks can be preformed and on their efficiency.

The field of information theory made great progress last century considering that the world is classical. Thanks to the seminal work by Turing,[1] it was possible to theoretically predict what problems could, in principle, be solved. This before the first calculator was actually built!

Last century we also learned that there is a more fundamental theory of nature: quantum mechanics. It then became interesting to revise information theory in the light of this new physical theory. This is what quantum information is. We have been learning to exploit quantum properties, such as entanglement, to improve information task. A good example of this is quantum teleportation.

However, something very important is missing in the picture. Las century we also learned that the world is not only quantum but also relativistic. In fact, before quantum information was conceived some relativistic considerations in information theory were considered: we all know that it is not possible to send messages faster than the speed of light.

The field of information theory is going through interesting times. The way we process information is being revolutionized by incorporating quantum theory. However, if we want to push this revolution further, as far as we possibly can, we need to incorporate relativity. It is important to do so now since most, if not all, implementations of quantum information in fact employ relativistic systems. Such is the case in Cavity QED and quantum information based protocols employing photons. The research of relativistic quantum field aims at understanding how to process information in the overlap of quantum theory and relativity: were real life experiments take place. And in the same way as we have been able to exploit quantum resources to improve information tasks we might be able to learn how to make use of relativistic effects as well.

2.2. *Abstract quantum information*

The main aim of quantum information theory is to learn how to store, process and read information using quantum systems. In this section we will briefly revise basic concepts of quantum information theory. Namely, quantum entanglement for pure and mixed states. Entanglement is a quantum property which is a consequence of the superposition principle and the tensor product structure of the Hilbert space. It plays a central role in the field of quantum information since it is at the heart of many quantum information tasks such as quantum teleportation and quantum cryptography.

2.2.1. *Pure states*

Entanglement for pure bi-partite systems is well understood. In what follows we will define and learn how to quantifying entanglement in this case.

In quantum theory the state of a quantum particle is a vector in a d-dimensional Hilbert space $\mathcal{H}$ where $\mathcal{H}$ is an inner product space over C. A state in the Hilbert space $\mathcal{H}$ is denoted $|\psi\rangle \in \mathcal{H}$. It is interesting to consider how we introduce a second particle to the description of our system. The state of two particles A and B is a vector in a $(d \times d')$-dimensional Hilbert space $\mathcal{H}_{ab} = \mathcal{H}_a \otimes \mathcal{H}_b$. The space $\mathcal{H}_{ab}$ is the tensor product of the sub-spaces $\mathcal{H}_a$ and $\mathcal{H}_b$ of each particle. An element of the space $\mathcal{H}_{ab}$ is written as $|\psi_{ab}\rangle = \sum_{i,j} A_{ij} |i\rangle_a \otimes |j\rangle_b$.

Note: This is very different to classical physics where if a particle has three degrees of freedom then two identical particles have $3 \oplus 3 = 6$ degrees of freedom where $\oplus$ denotes the direct sum. In quantum mechanics the vector state describing the state of the two 3-dimensional particles is $3 \times 3 = 9$ dimensional.

Given the tensor product structure of the Hilbert space for two particles we can consider the following definitions:

Definition 2.1. A state $|\psi_{ab}\rangle \in \mathcal{H}_a \otimes \mathcal{H}_b$ is seperable if $|\psi_{ab}\rangle = |\psi\rangle_a \otimes |\psi\rangle_b$.

Comments (physics)

(1) A separable state can be prepared by local operations and classical communication. This means that observers manipulate each particle independently by making measurements or applying unitary transformations of the form $|\psi_{ab}\rangle = \hat{U}_a \otimes \hat{U}_b |\psi_i\rangle$ where U_a and U_b are unitaries acting on particle A and B, respectively. The observers are also allowed to exchange classical information.

(2) By measurements on particle A one learns nothing about particle B.

Definition 2.2. If the state is not seperable then it is entangled.

Comments (physics):

(1) An entangled state cannot be prepared by local operations and classical communication. Observers must make global operations on the systems. An example are interactions between A and B.

(2) By measurements on particle A one can learn infomation about B.

Example 2.1. Two qubits

A state $|\psi\rangle = \frac{1}{\sqrt{2}}(|0\rangle_a |1\rangle_b + |1\rangle_a |1\rangle_b)$ in the Hilbert space $\mathcal{H}^2 \otimes \mathcal{H}^2$ can be written as the product

$$|\psi\rangle = \frac{1}{\sqrt{2}}(|0\rangle_a + |1\rangle_a) \otimes |1\rangle_b$$

and is thus, a seperable state. If we measure subsystem B then we cannot learn anything about subsystem A.

The state $|\phi\rangle = \frac{1}{\sqrt{2}}(|0\rangle_a |0\rangle_b + |1\rangle_a |1\rangle_b)$ is not a seperable state since we cannot write it as the product of two states that belong seperatlety to the individual subsytems $\mathcal{H}_a$ and $\mathcal{H}_b$ i.e.

$$|\phi\rangle \neq |\psi_a\rangle \otimes |\psi_b\rangle$$

Question: How do we know if a general state $|\psi_{ab}\rangle = \sum A_{ij} |i\rangle_a |j\rangle_b$ is entangled or not? To answer this consider the following theorem:

Theorem 2.1. *Let $\mathcal{H}_1$ and $\mathcal{H}_2$ be d-dimensional Hilbert spaces. For any vector $|\psi_{ab}\rangle \in \mathcal{H}_a \otimes \mathcal{H}_b$ there exists a set of orthonormal vectors*

$$\{|j\rangle_a\} \subset \mathcal{H}_a \text{ and } \{|l\rangle_b\} \subset \mathcal{H}_b$$

such that we can write

$$|\psi_{ab}\rangle = \sum_i \lambda_i |i\rangle_a |i\rangle_b$$

where the Schmidt coefficients λ_i are non-negative scalars.

This special basis is called the Schmidt basis. For simplicity we assumed the Hilbert spaces for each system to have the same dimension. However, this statement can be generalized to a $d \times d'$ system. Note that the correlations between systems A and B are now made explicit. Therefore, if $\lambda_{i \neq j} = 0$ and $\lambda_j = 1$ then the state is seperable. In the case that all the λ_i's are equal and $\lambda_i = \frac{1}{\sqrt{d}}$ then the state is maximally entangled.

It is clear that the distribution of the Schmidt coefficients determine how entangled the state is. Therefore, to quantify entanglement in the pure bi-partite case we need a monotonous and continuous function of the λ_i's such that

    (1) $S(\lambda_i) = 0$ for separable states
    (2) $S(\lambda_i) = \log(d)$ for maximally entangled states

Considering the density matrix $\rho_{ab} = |\psi_{ab}\rangle\langle\psi_{ab}|$ and it's reduced density matrix $\rho_b = Tr_a(\rho_{ab})$, we find that the Von-Neuman entropy

$$S(\rho_b) = -Tr(\rho_b \log_2 \rho_b)$$
$$= -\sum_i |\lambda_i|^2 \log_2 |\lambda_i|^2$$

quantifies the entanglement between system A and B. We observe from the Schmidt decomposition that it is equivalent to trace over either system A or B and therefore, $S(\rho_a) = S(\rho_b)$. The Von-Neuman entropy of a pure state is $S(\rho_{ab}) = 0$.

Example 2.2. Calculate the Von-Neumann entropy of the state $|\psi\rangle_{ab} = \frac{1}{\sqrt{2}}(|0,0\rangle + |1,1\rangle)$. The density matrix is given by

$$\rho_{ab} = \frac{1}{2}(|0,0\rangle\langle0,0| + |0,0\rangle\langle1,1| + |1,1\rangle\langle0,0| + |1,1\rangle\langle0,0| + |1,1\rangle\langle1,1|) \quad (1)$$

Tracing over the first system we obtain

$$\rho_a = \frac{1}{2}(|0\rangle\langle0| + |1\rangle\langle1|) \quad (2)$$

which is already in diagonal form. The matrix has two degenerate eigenvalues equal to $1/2$. Therefore, the Von-Neumann entropy is $S(\rho_a) = 2(1/2\log_2(1/2)) = 1$ which shows that the state is maximally entangled.

2.2.2. *Mixed states*

Quantifying entanglement in the mixed case is more involved since there is no analog to the Schmidt decomposition in this case. However, it is possible to define what a separable mixed state is. A mixed state is separable if we can write its density matrix as

$$\rho_{ab} = \sum_i \omega_i \rho_a^i \otimes \rho_b^i$$

where $\sum_i \omega_i = 1$.

To find out if a general mixed state is entangled or not it is convenient to define the partial transpose of a density matrix. Consider the general mixed state

$$\rho_{ab} = \sum_{ijkl} C_{ijkl}|i\rangle_a|j\rangle_{bb}\langle k|_a\langle l|. \quad (3)$$

The partial transpose ρ_{ab}^{PT} of ρ_{ab} is

$$\rho_{ab}^{PT} = \sum_{ijkl} C_{ljki}|i\rangle_a|j\rangle_{bb}\langle k|_a\langle l|. \quad (4)$$

or equivalentely

$$\rho_{ab}^{PT} = \sum_{ijkl} C_{ikjl} |i\rangle_a |j\rangle_{bb}\langle k|_a\langle l|. \tag{5}$$

Since the partial transpose of a separable state has positive eigenvalues it is possible to construct a separability criterion.

Necessary Condition for Seperability (Peres): If the eigenvalues of $\rho_{ab}^{PT} \geqslant 0$ then ρ_{ab} is separable.

However, this criterion is only sufficient for 2×2 and 2×3 systems. For systems of higher dimension the criterion is only necessary meaning that there are entangled states with positive partial transpose. Such states are known as bound entangled states.

Adding up the negative eigenvalues gives an estimate of how entangled a state is. Therefore, we will now define the negativity and logarithmic negativity which are two entanglement monotones.

Definition 2.3. The negativity of a density matrix ρ_{ab} is defined as the sum of the negative eigenvalues of the partial transpose ρ_{ab}^{PT}

$$N(\rho_{ab}) := \frac{\|\rho_{ab}^{PT}\| - 1}{2}$$

Where $\| \cdot \|$ denotes the trace norm $\|X\| := Tr\left[\sqrt{X^\dagger X}\right]$.

Definition 2.4. The logarithmic negativity of a density matrix ρ_{ab} is defined as

$$E_N(\rho_{ab}) := \log_2 \|\rho_{ab}^{PT}\|$$

REFERENCES

- For a review on entanglement: Dagmar Bruss, J. Math. Phys., (43): 4237, 2002.
- On the negativity monotone: G. Vidal and R. F. Werner, Phys. Rev. A, (65): 032314, 2002.

3. Technical Tools in Relativity

We are interested in understanding entanglement when the underlying spacetime is considered in the description of states. In this lecture we will learn some basic notions of spacetime which is a 4-dimentional manifold $\mathcal{M}$ (space) with a Lorentizian metric. For our purpose it is enough to consider that $\mathcal{M}$ is a collection of points which locally looks like $\mathbb{R}^4$. We will introduce the basic mathematical structures we need to define a Lorentzian metric, a light cone and a world line. We will start with the following (strong) remarks:

There is <u>no</u> time
There is <u>no</u> space
There are only cones.

3.1. *Necessary mathematical structures*

In order to clarify these remarks we will introduce the following mathematical structures:

(1) **Coordinates** are 4-functions χ^μ such that $\chi^\mu : p \in M_a \to \mathbb{R}$ where p is a point in the region $\mathcal{M}_a$ of the manifold $\mathcal{M}$. Since $\mu = \{0, 1, 2, 3\}$, the functions $\chi^\mu = (\chi^0, \chi^1, \chi^2, \chi^3)$. A choice of coordinates is not unique, therefore, it is possible to choose other coordinates in M_a, $\bar{\chi}^\mu = \bar{\chi}^\mu(\chi^0, \chi^1, \chi^2, \chi^3)$ such that

$$
\begin{aligned}
d\bar{\chi}^\mu &= \frac{\partial \bar{\chi}^0}{\partial \chi^0}d\chi^0 + \frac{\partial \bar{\chi}^1}{\partial \chi^1}d\chi^1 + \frac{\partial \bar{\chi}^2}{\partial \chi^2}d\chi^2 + \frac{\partial \bar{\chi}^3}{\partial \chi^3}d\chi^3 \\
&= \frac{\partial \bar{\chi}^\mu}{\partial \chi^\nu}d\chi^\nu
\end{aligned}
$$

Note that we sum over the repeated indices (Einstein's convention). The transformation is well defined were the Jacobian is non-zero i.e.

$$
det\left(\frac{\partial \bar{\chi}^\mu}{\partial \chi^\nu}\right) \neq 0
$$

(2) **Contravariant vectors** are a set of quantities that transform according to

$$
\bar{\chi}^\mu = \frac{\partial \bar{\chi}^\mu}{\partial \chi^\nu}\chi^\nu \tag{6}
$$

where $\bar{\chi}^\mu$ and χ^ν are coordinates associated with the point p. In the same fashion we define contravariant tensors as the set of quantities that transform as

$$\bar{T}^{\mu\nu} = \frac{\partial \bar{\chi}^\mu}{\partial \chi^\alpha} \frac{\partial \bar{\chi}^\nu}{\partial \chi^\beta} T^{\alpha\beta} \tag{7}$$

(3) **Covariant tensors** transform according to

$$\bar{T}_{ab} = \frac{\partial \chi^c}{\partial \bar{\chi}^a} \frac{\partial \chi^d}{\partial \bar{\chi}^b} T_{cd}$$

We say that quantities with a single index are of rank 1 (vectors) and that quantities that have two indices are of rank 2.

(4) **Metric tensors** are rank 2 covariant tensors which are symmetric i.e.

$$g_{\alpha\beta} = g_{\beta\alpha}$$

If $det(g) \neq 0$ then the metric is non-singular and we can define the inverse metric tensor $g^{\alpha\beta}$ such that

$$g_{ab}g^{bd} = \delta_a{}^d.$$

3.2. *Lorentzian metrics, light cones and world lines*

We are now in position to define the basic notions of spacetime.

(1) A **Lorentzian metric** is a metric with signature $\{+ - - -\}$. That is if at any given point in spacetime we can find coordinates such that

$$g_{ab} = \eta_{ab} = \begin{pmatrix} 1 & & & \\ & -1 & & \\ & & -1 & \\ & & & -1 \end{pmatrix}$$

Example 3.1. In flat spacetime the metric $g_{ab} = \eta_{ab}$ everywhere.

Metrics are used to define distances and length vectors.

(2) A **line element** is the infinitesimal distance between two neighbouring points χ^a and $\chi^a + d\chi^a$ defined as

$$ds^2 = g_{ab}(\chi)d\chi^a d\chi^b$$

.

Example 3.2. In flat spacetime in $(1+1)$-dim the metric is $\eta_{ab} = \{+-\}$ and so

$$\begin{aligned} ds^2 &= \eta_{ab}d\chi^a d\chi^b \\ &= \eta_{00}(d\chi^0)^2 + \eta_{11}(d\chi^1)^2 \\ &= dt^2 - dx^2 \end{aligned}$$

(3) The **norm of a contravariant vector** χ^a is defined as

$$\begin{aligned} g(\chi,\chi) &:= \chi^2 \\ &= g_{ab}\chi^a\chi^b \end{aligned}$$

We can therefore define three types of vector as

   (a) timelike $g_{ab}\chi^a\chi^b > 0$
   (b) spacelike $g_{ab}\chi^a\chi^b < 0$
   (c) null or lightlike $g_{ab}\chi^a\chi^b = 0$

(4) **Light cone** The set of all null vectors at p define a light cone. $\eta_{ab}\chi^a\chi^b = 0$ is the equation of a double light cone.

(5) A **timelike world line** is a curve $\chi^\mu(u)$ whose tangent vector is everywhere timelike. These are the tracks where particles and observers can travel.

Using the definition of the line element we can derive the interval between two spacetime points p_1 and p_2. Taking the line element and dividing by du^2 we see that

$$\left(\frac{ds}{du}\right)^2 = g_{ab}\left(\frac{d\chi^a}{du}\right)\left(\frac{d\chi^b}{du}\right)$$

where u parametrizes the trajectory. Thus integrating over u we find the proper distance between two points is

$$L[\chi] := \int \sqrt{g_{ab}\left(\frac{d\chi^a}{du}\right)\left(\frac{d\chi^b}{du}\right)}\,du$$

REMARK This implies that there is no global time and no global space. There is a well defined notion of time for each observer given by the length $L[\chi] = \zeta$.

Example 3.3. We will find the line element ds^2 of the $(1+1)$-dim flat spacetime in the following coordinates
a) $u = t - x$ and $v = t + x$
b) $x = \frac{e^{a\xi}}{a}\cosh(a\zeta)$ and $t = \frac{e^{a\xi}}{a}\sinh(a\zeta)$ where a is a constant. These coordinates are known as Rindler coordinates.

a) The line element

$$ds^2 = g_{ab}(X)dX^a dX^b$$

in flat spacetime is given by $ds^2 = dt^2 - dx^2$. Thus, to find the line element in the new coordinates we need to calculate the derivatives

$$du = dt - dx$$
$$dv = dt + dx$$

and thus

$$dt = \frac{1}{2}(du + dx)$$
$$dx = \frac{1}{2}(du - dx)$$

therefore,

$$\begin{aligned} ds^2 &= dt^2 - dx^2 \\ &= \frac{1}{4}((du + dv)^2 - (du - dv)^2) \\ &= \frac{1}{4}((du^2 + dv^2 + 2dudv) - (du^2 + dv^2 - 2dudv)) \\ &= dudv \end{aligned}$$

b) Given the coordiates $x = \frac{1}{a}e^{a\eta}\sinh(a\zeta)$ and $t = \frac{1}{a}e^{a\eta}\cosh(a\zeta)$, where a is a constant, we find that

$$dt = d\eta e^{a\eta}\cosh(a\zeta) + d\zeta e^{a\eta}\sinh a\zeta$$
$$dx = d\eta e^{a\eta}\sinh(a\zeta) + d\zeta e^{a\eta}\cosh a\zeta$$

therefore,

$$\begin{aligned} ds^2 &= dt^2 - dx^2 \\ &= e^{2a\eta}(\cosh(a\zeta)d\eta + \sinh(a\zeta)d\zeta)^2 - e^{2a\eta}(\sinh(a\zeta)d\eta + \cosh(a\zeta)d\zeta)^2 \\ &= e^{2a\eta}(\cosh^2(a\zeta) - \sinh^2(a\zeta))d\eta^2 + e^{2a\eta}(\sinh^2(a\zeta) - \cosh^2(a\zeta))d\zeta^2 \\ &= e^{2a\eta}(d\eta^2 - d\zeta^2) \end{aligned}$$

Example 3.4. In a black hole spacetime the line element is given by

$$ds^2 = (1 - (2m/R))dT^2 - (1 - (2m/R))^{-1}dR^2$$

where m is the mass of the black hole and R the radius. Note how, far from the Schwarzschild radius $R = 2m$, spacetime is flat. As we come closer to the Schwarzschild radius the light cones tilt. At the Schwarzschild radius the light cones are tilted such that particles following world lines can fall into the black hole but cannot escape.

REFERENCES

Introducing Einstein's relativity by Ray d'Inverno. Published 1995 by Clarendon Press, Oxford University Press in Oxford New York .

4. Technical Tools from Quantum Field Theory I

We are interested in understanding quantum information in a relativistic quantum world. For this we must consider the most general situation: quantum field theory in curved spacetime. The most important lesson we have learned from quantum field theory is that fields are fundamental notions, and not particles. As we will see, particles are derived notions (if at all possible). In this lecture we will therefore consider fields on a spacetime. We will learn how to quantize a field in curved spacetime and under what circumstances particles can be defined. This is important for information theory in which all concepts are based on the notions of subsystems (particles). Throught our lectures we will work in natural units $c = \hbar = 1$.

4.1. *The Klein-Gordon equation*

We will consider the simplest field which is the Klein-Gordon un-charged scalar field ϕ. The field ϕ satisfies the Klein-Gordon equation $(\Box + m^2)\phi = 0$, where the operator $\Box$ is known as the d'Alambertian and is defined as

$$\Box\phi := \frac{1}{\sqrt{-g}}\partial_\mu(\sqrt{-g}g^{\mu\nu}\partial_\nu\phi)$$

where $g = det(g^{ab})$ and $\partial_\mu = \frac{\partial}{\partial\chi^\mu}$.

Example 4.1. In flat spacetime $(1+1)$-dim we have $g_{\nu\mu} = \eta_{\nu\mu} = \{+-\}$ and thus

$$\Box\phi = \partial_0(\eta^{00}\partial_0) + \partial_1(\eta_1^1\partial_1)$$
$$= \partial_t^2 - \partial_x^2$$

For $m = 0$ the solutions to $\Box\phi = 0$ in 2-dim are plane waves of the form

$$u_k = \frac{1}{\sqrt{2\pi\omega}}e^{i(kx-wt)}$$

with $\omega = |k|$ and $-\infty < k < \infty$.

We would like to quantize this field. For this we note that the solutions of the equation form a vector space over $\mathbb{C}$. We can therefore, consider the following (bad!) idea:

By supplying an inner product we can construct a Hilbert space and possibly consider $\hat{u}_k$ as operators. However, the inner product must be Lorentz invariant. The only possible choice is

$$(\phi, \psi) = -i \int_\Sigma (\psi^* \partial \mu \phi - (\partial_\mu \psi^*)\phi) d\Sigma^\mu$$

where Σ is a spacelike hypersurface.

We then note that (ϕ, ψ) can be negative! Therefore, we cannot define a probability density. This means that it is not possible to construct a one-particle relativistic quantum theory.

4.2. *Time-like Killing vector fields*

To understand the deeper reasons why this naive program does not work, let us introduce the notion of a time-like Killing vector field. We are interested in finding a transformation $\chi^a \to \bar{\chi}^a$ which leaves the metric $g_{ab}(\chi)$ invariant. That means that the transformed metric $\bar{g}^{ab}(\bar{x})$ is the same function of its argument $\bar{\chi}^a$ as the original metric $g_{ab}(\chi)$ is of its argument χ^a. Since the metric is a covariant tensor it transforms according to

$$g_{ab}(\chi) = \frac{\partial \bar{\chi}^c}{\partial \chi^a} \frac{\partial \bar{\chi}^d}{\partial \chi^b} \bar{g}_{cd}(\bar{\chi})$$

Therefore, the metric is invariant under the transformation $\chi^a \to \bar{\chi}^a$ if

$$g_{ab}(\chi) = \frac{\partial \bar{\chi}^c}{\partial \chi^a} \frac{\partial \bar{\chi}^d}{\partial \chi^b} g_{cd}(\bar{\chi}).$$

This equation is complicated therefore, it is easier to consider the infinitesimal transformation $\chi^a \to \chi^a + \delta u X^a(\chi) = \bar{\chi}^a$ where δu is small and arbitrary and X is a vector field. Therefore we ask that in the limit $\delta u \to 0$

$$\lim_{\delta u \to 0} \frac{g_{ab}(\bar{\chi}) - \bar{g}_{ab}(\bar{\chi})}{\delta u} = 0.$$

Differentiating $\bar{\chi}^a = \chi^a + \delta u X^a(\chi)$

$$\frac{\partial \bar{\chi}^a}{\partial \chi^c} = \delta^a{}_c + \delta u \partial_c X^a$$

$$\frac{\partial \bar{\chi}^b}{\partial \chi^d} = \delta^b{}_d + \delta u \partial_d X^b$$

We find that in a Taylor expansion of $g_{ab}(\bar{\chi})$

$$g_{ab}(\bar{\chi}) = g_{ab}(\chi^e + \delta u X^e) = g_{ab}(\chi) + \delta u X^e \partial_e g_{ab}(\chi) + ...) \qquad (8)$$

therefore,

$$\begin{aligned}
g_{ab}(\chi) &= (\delta^c_{\ a} + \delta u \partial_a X^c)(\delta^d_{\ b} + \delta u \partial_d X^b)(g_{cd}(\chi^e) + X^e \delta u \partial_e g_{cd}(\chi^e) + ...) \\
&= \delta^c_{\ a} \delta^d_{\ b} g_{cd}(\chi) + \delta u \left[\delta^c_{\ a} \partial_d X^b g_{cd}(\chi) + \delta^d_{\ b} \partial_a X^c g_{cd}(\chi) + X^e \partial_e g_{cd}(\chi) \right] \\
&\quad + \theta(du^2) \\
&= g_{ab}(\chi) + \delta u \left[X^e \partial_e g_{ab}(\chi) + g_{ad} \partial_b X^d + g_{bd} \partial_a X^d \right] + \theta(du^2)
\end{aligned}$$

Working to first order in δu and subtracting $g_{ab}(\chi)$ on both sides, it follows that the quantity in the bracket must vanish,

$$\left[X^e \partial_e g_{ab}(\chi) + g_{ad} \partial_b X^d + g_{bd} \partial_a X^d \right] = 0$$

We identify the object inside the bracket with the Lie deriviative of the metric tensor

$$\mathcal{L}_X g_{ab} = X^e \partial_e g_{ab}(\chi) + g_{ad} \partial_b X^d + g_{bd} \partial_a X^d$$

Therefore, we say that X is a Killing vector field if the equation $\mathcal{L}_X g^{ab} = 0$ is satisfied.

4.3. *Quantizing the field and the definition of particles*

Having defined a Killing vector field X, let us choose a special basis for the solutions of $\Box \phi = 0$ such that

$$i\mathcal{L}_X u_k = iX^\mu \partial_\mu u_k$$
$$= \omega u_k$$

where we have considered the action of a Lie derivative on a function. Vectors lying within the light cone at each point are called time-like. Therefore, if X is a timelike vector field, the Lie deriviative corresponds to ∂_t. This implies that the eigenvalue equation above takes form of a Schrödinger-like equation where we can identify $\omega > 0$ with a frequency

$$i\partial_t u_k = -i\omega u_k$$
$$i\partial_t u_k^* = i\omega u_k^*$$

The solutions to the Klein-Gordon equation are therefore classified in the following way

$$u_k \rightarrow \text{positive frequence solutions}$$
$$u_k^* \rightarrow \text{negative frequence solutions}$$

Interestingly, we observe that the inner product is positive for positive frequency solutions and negative for negative frequency solutions

$$(u_k, u_{k'}) \geqslant 0$$
$$(u_k^*, u_{k'}^*) \leqslant 0$$

We found the culprits! Negative frequency solutions give rise to negative probabilities. However, the real problem with trying to construct a single-particle quantum field theory is the following: we know from experiment that in field theory there are multi-particle interactions.

The single-particle Hilbert space in not big enough nor appropriate to construct a quantum field theory. We need a bigger space. The right mathematical structure is given by a Fock space. The Fock space is constructed in the following way: we consider a zero-particle sector which contains the vacuum state $|0\rangle$. Then a single particle Hilbert space is constructed as proposed before. We use the vector field of solutions of the equation and impose the Lorentz invariant inner product. We now allow for multi-particle sectors which are constructed using the tensor product structure. For example, the Hilbert space for a 2-particle sector is given by $\mathcal{H}_2 = \mathcal{H} \otimes \mathcal{H}$. The Fock space therefore takes the form

$$\mathbb{C} \oplus \mathcal{H} \otimes \mathcal{H} \oplus \mathcal{H} \otimes \mathcal{H} \otimes \mathcal{H} \oplus \mathcal{H} \otimes \mathcal{H} \otimes \mathcal{H} \otimes \mathcal{H} \otimes ...$$

It is then possible to define creation and annihilation operators $a_k^\dagger$ and a_k which, in the case of bosonic fields, satisfy the commutation relations $[a_k^\dagger, a_{k'}] = \delta_{k,k'}$. Creation operators $a_k^\dagger$ take us from the n-particle sector to the (n+1)-particle sector by creating a particle of momentum k. Annihilation operators a_k take us from the n-particle sector to the (n-1)-particle sector by annihilating a particle with momentum k. Having defined creation and annihilation operators, we can define the following operator value function being careful to treat the positive and negative solutions of the Klein-Gordon equation in a different way

$$\hat{\phi} = \int (u_k a_k + u_k^* a_k^\dagger) dk.$$

Notice that positive frequency solutions are associated with annihilation operators and negative frequency solutions with creation operators. Amazingly, this operator value function satisfies the Klein-Gordon equation $\Box \hat{\phi} = 0$. This is in fact, the right way to quantize a field. Now, we can properly define particles as the action of creation operators on the vacuum state

$$|n_1, ..., n_k'\rangle = a_1^{\dagger n}...a_k^{\dagger n'}|0\rangle$$

and such states have positive norm.

REMARKS When there exists a global time-like Killing vector field it is meaningful to define particles. Observers flowing along timelike Killing vector fields are those who can properly describe particle states. This has important consequences to relativistic quantum information since the notion of particles (and therefore, subsystems) are indispensable to store information and thus, to define entanglement.

REFERENCES

- *Introducing Einstein's Relativity* by Ray d'Inverno. Published 1995 by Clarendon Press, Oxford University Press in Oxford New York .

- *Quantum Field Theory in Curved Spacetime* by N.D. Birrell and P.C.W. Davies, CUP (1982).

5. Technical Tools in Quantum Field Theory II

In the last lecture we learned that only observers flowing along timelike Killing vector fields can meaningfully describe particles. However, in the most general case, curved spacetimes do not admit such structures. There are in fact different kind of spacetimes

(1) Spacetimes with global Killing vector fields
(2) Spacetimes with no Killing vector fields
(3) Spacetimes with regions which admit Killing vector fields

In this lecture we will see that in the case that the spacetime admist a timelike global Killing vector field, the vector field is not necessarily unique. A consequence of this is that the particle content of the field is observer-dependent. In the following lectures we will analyze the consequences of this for entanglement in spacetime. As an example of the observer-dependent property of the field, we will consider observers in flat-spacetime and derive the Unruh effect.

5.1. *Killing observers and Bogolubov transformations*

When a spacetime admits a timelike Killing vector field ∂_t it is possible to classify solutions to the Klein-Gordon equation $\{u_k, u_k^*\}$ into positive and negative frequency solutions. The field is therefore quantized as

$$\hat{\phi} = \int (u_k a_k + u_k^* a_k^\dagger) dk.$$

However, when such is the case, ∂_t is not generally unique. It is possible to find another time-like Killing vector field $\partial_{\bar{t}}$ and therefore find another basis for the solutions to the Klein-Gordon equation $\{\bar{u}_k, \bar{u}_k^*\}$ such that classification into positive and frequency solutions is possible. The field then is equivalently quantized in this basis as

$$\hat{\phi} = \int (\bar{u}_{k'} \bar{a}_{k'} + \bar{u}_{k'}^* \bar{a}_{k'}^\dagger) dk'.$$

and therefore,

$$\hat{\phi} = \int (u_k a_k + u_k^* a_k^\dagger) dk = \int (\bar{u}_{k'} \bar{a}_{k'} + \bar{u}_{k'}^* \bar{a}_{k'}^\dagger) dk'.$$

Using the inner product, it is then possible to find a transformation between them the creation and annihilation operators

$$a_k = \sum_{k'} (\alpha_{kk'}^* \bar{a}_{k'} - \beta_{kk'}^* \bar{a}_{k'}^\dagger)$$

where $\alpha_{kk'} = (u_k, \bar{u}_{k'})$ and $\beta_{kk'} = -(u_k, \bar{u}_{k'}^*)$ are called Bogolubov coefficients. Since the vacua states are defined as

$$a_k |0\rangle = \bar{a}_k |\bar{0}\rangle = 0$$

it is possible to find a transformation between the states in the two basis. We note that as long as one of the Bogoluvob coefficients $\beta_{kk'}$ is non-zero, while the un-barred state is the vacuum state, the state in the bared basis is populated with particles.

REMARK Different Killing observers observe a different particle content in the field. Therefore, particles are observer-dependent quantities. As an example we will analyze observers in flat spacetime.

5.2. *Example 1: Observers in flat spacetime*

Consider the scalar field $\phi(t, x)$ defined in all points of Minkowski spacetime in 1+1 dimensions. The line element is given by $ds^2 = dt^2 - dx^2$. The Killing vector fields in this spacetime are given by the equation

$$\mathcal{L}_X \eta_{ab} = 0$$

which in components takes the form

$$X^e \partial_e \eta_{ab} + \eta_{ad} \partial_b X^d + \eta_{bd} \partial_a X^d = 0.$$

In flat spacetime there are three independent Killing vector fields corresponding to

(1) Time translationss
(2) Space translation
(3) Observers moving on hyperbolas

To show this we use the Killing equation shown above finding that

$$\partial_t X^0 = \partial_x X^1 \Rightarrow X^\mu = \alpha \begin{pmatrix} t \\ x \end{pmatrix}$$

where α is a constant. The condition for this Killing vector field to be timelike is

$$0 \leqslant \eta_{ab} X^a X^b \Rightarrow 0 \leqslant \alpha^2 (t^2 - x^2).$$

Possible solutions correspond to the case where t and x are constants giving rise to

$$X^\mu = \alpha t \begin{pmatrix} 1 \\ 0 \end{pmatrix} + \alpha x \begin{pmatrix} 0 \\ 1 \end{pmatrix}$$

where the first summand corresponds to time translations and the second to space translations. Therefore, from time and space translations we obtain straight lines corresponding to inertial observers. Another possible solution are hyperbolas $t^2 - x^2 = const$. In what follows we will show that observers moving along these trajectories correspond to observers in uniform acceleration.

In relativity acceleration is uniform if, at each instant, the acceleration in an inertial frame travelling with the same velocity as the particle has the same value. We therefore consider two observers: the first with coordinates (t, x) and the second, who is inertial, with coordinates $(\tilde{t}, \tilde{x})$ moving with constant velocity v with respect to the first . Using the Lorentz transformations

$$\tilde{t} = \gamma \left(t - \frac{vx}{c^2} \right), \quad \tilde{x} = \gamma (x - vt)$$

where $\gamma = (1 - \frac{v^2}{c^2})^{-1}$ is the Lorentz factor, we aim at finding the transformation between the accelerations in the two reference frames

$$a = \ddot{x} = \frac{d^2 x}{dt^2}, \quad \tilde{a} = \ddot{\tilde{x}} = \frac{d^2 \tilde{x}}{d\tilde{t}^2}.$$

In our notation dot represents differentiation with respect to the time coordinate. Since $d\tilde{x}$ and $d\tilde{t}$ are given by

$$d\tilde{x} = \gamma (dx - vdt), \quad c^2 d\tilde{t} = \gamma (c^2 dt - vdx)$$

we obtain the following transformation between the velocities

$$\frac{\tilde{\dot{x}}}{c^2} = \frac{1}{c^2}\frac{d\tilde{x}}{d\tilde{t}} = \frac{\gamma(dx - vdt)}{\gamma(c^2dt - vdx)} = \frac{(dx - vdt)}{(c^2dt - vdx)} = \frac{(\dot{x} - v)}{(c^2 - v\dot{x})}.$$

Denoting $\tilde{\dot{x}} \equiv \tilde{u}$ and $\dot{x} \equiv u$ and differentiating yields

$$c^{-2}d\tilde{u} = d[u - v](c^2 - vu)^{-1} + (u - v)d[c^2 - vu]^{-1}$$
$$= du(c^2 - vu)^{-1} + v(u - v)(c^2 - vu)^{-2}du$$
$$= (c^2 - vu)^{-2}[c^2 - vu + v(u - v)]du$$
$$= (c^2 - vu)^{-2}(c^2 - v^2)du.$$

Hence

$$\frac{d\tilde{u}}{d\tilde{t}} = c^4 \frac{(c^2 - vu)^{-2}(c^2 - v^2)}{\gamma(c^2 - vu)}\frac{du}{dt}$$
$$\tilde{a} = c^3(c^2 - vu)^{-3}(c^2 - vu)^{\frac{3}{2}}a$$

which yields the transformation between accelerations we were looking for

$$\tilde{a} = \left(1 - \frac{uv}{c^2}\right)^{-3}\left(1 - \frac{v^2}{c^2}\right)^{\frac{3}{2}}a.$$

The inverse transformation yields,

$$a = \tilde{a}\left(1 - \frac{v^2}{c^2}\right)^{\frac{3}{2}}\left(1 + \frac{\tilde{u}v}{c^2}\right)^{-3}.$$

To consider an observer moving in uniform acceleration we set $\tilde{a} = \alpha$ where α is a constant. We also must consider that the inertial observer moves at each time with the same velocity as the first observer. Therefore, $\tilde{u} = 0$ and $v = u$ which results in

$$\frac{d^2x}{dt^2} = \alpha\left(1 - \frac{1}{c^2}\left(\frac{dx}{dt}\right)\right)^{\frac{3}{2}}.$$

To find the trajectory of the observer in uniform acceleration we integrate the above equation

$$\frac{du}{\left(1 - \frac{u^2}{c^2}\right)^{\frac{3}{2}}} = \alpha dt$$

and assuming the particle starts from rest at time $t = t_0$, we find

$$\frac{u}{\left(1 - \frac{u^2}{c^2}\right)^{\frac{1}{2}}} = \alpha(t - t_0)$$

and integrating once more we obtain,

$$(x - x_0) = \frac{c}{\alpha}[c^2 + \alpha^2(t - t_0)^2]^{\frac{1}{2}} - \frac{c^2}{\alpha}$$

or equivalently,

$$\frac{(x - x_0 + c^2/\alpha)^2}{(c^2/\alpha)} - \frac{(ct - ct_0)^2}{(c^2/\alpha)} = 1.$$

Setting $x_0 - (c^2/\alpha) = t_0 = 0$, we obtain $x^2 - c^2t^2 = c^2/\alpha$. Therefore, we have found that an observer in uniform acceleration moves along hyperbolas. In the next lecture we will investigate how the field looks like from the perspective of an observer in uniform acceleration given that the field is in the vacuum from the inertial perspective.

REFERENCES

- *Introducing Einstein's relativity* by Ray d'Inverno. Published 1995 by Clarendon Press, Oxford University Press in Oxford New York.
- *Quantum Field Theory in Curved Spacetime* by N.D. Birrell and P.C.W. Davies, CUP (1982).

6. The Unruh Effect

In this lecture we continue with our example of observers in flat spacetime. We have shown that in flat spacetime there are two kinds of observers who can meaningfully descibe particles: inertial observers who's trajectories correspond to straight lines and observers in uniform acceleration who follow hyperbolas. We have learned that, in principle, the particle content of the field can be different when described from the perspective of different observers. Therefore, we will quantize the field from the perspective of inertial observers and then describe it from the perspective of observers in uniform acceleration.

Minkowski coordinates (t, x) are a convenient choice of coordinates for inertial observers. We saw in our last lecture that their trajectories follow straight lines. In this coordinates the Klein-Gordon equation takes the following form

$$(\partial_t^2 - \partial_x^2)\phi = 0.$$

The solutions to this equation are plane waves

$$u_k = \frac{1}{\sqrt{2\pi\omega}} e^{i(kx-\omega t)}$$

$$u_k^* = \frac{1}{\sqrt{2\pi\omega}} e^{-i(kx-\omega t)}$$

with $\omega = |k|$ and $-\infty < k < \infty$. The spacetime admits global timelike Killing vector fields. In this case, the timelike Killing vector field corresponds to ∂_t. With respect to this field, we can classify the above solutions in positive and negative frequency solutions. Since

$$i\partial_t u_k = -i\omega u_k$$

$$i\partial_t u_k^* = i\omega u_k^*$$

we identify u_k as positive and u_k^* as negative frequency solutions. Since these solutions form a complete set of orthonormal functions in Minkowski spacetime the quantized field can be expressed as

$$\hat{\phi} = \int (u_k a_k + u_k^* a_k^\dagger) dk$$

were the creation and annihilation operators follow the appropriate commutation relations $[a_k^\dagger, a_{k'}] = \delta_{k,k'}$. The vacuum state from this perspective is therefore defined as $a_k |0\rangle^{\mathcal{M}} = 0$. The state can be written as $|0\rangle^{\mathcal{M}} = \prod_k |0_k\rangle^{\mathcal{M}}$ where $|0_k\rangle^{\mathcal{M}}$ is the vacuum state of mode k.

Now we want to describe the field from the perspective of an observer moving with uniform acceleration. The trajectory of such observer follows a hyperbola. We will parametrize the trajectory in the following way

$$x = \frac{e^{a\chi}}{a} \cosh(a\eta)$$

$$t = \frac{e^{a\chi}}{a} \sinh(a\eta)$$

such that $x^2 - t^2 = \frac{e^{a\chi}}{a}$. Here a is an arbitrary reference acceleration. The observer's proper acceleration is given by $ae^{-a\chi}$. This suggests that appropriate coordinates for accelerated observers are (η, χ) which are know as Rindler coordinates. The temporal coordinates $\eta = cte$ correspond to straight lines going through the origin and the spacial coordinates $\chi = cte$ are hyperbolas. Note that in the limit that $\chi \to -\infty$ we have $x^2 = t^2$ which corresponds to lines at 45 degrees. The transformation is defined in the region $|x| \geq t$ which defines a wedge known as Rindler wedge I. When $\eta \to \infty$ then

$$\frac{x}{t} = \tanh(a\eta) \to 1 \Rightarrow x = t$$

Observers uniformly accelerated asymptotically approach the speed of light and are constrained to move in wedge I. Since the transformation does not cover the whole Minkowski space, we must define a second region called Rindler wedge II by considering the coordinate transformation

$$x = \frac{-e^{a\chi}}{a} \cosh(a\eta)$$
$$t = \frac{-e^{a\chi}}{a} \sinh(a\eta)$$

which differ from the first set of transformations by a sign in both coordinates. Rindler regions I and II are causally disconnected and the lines at 45 degrees define the Rindler horizon. No information can flow between these regions.

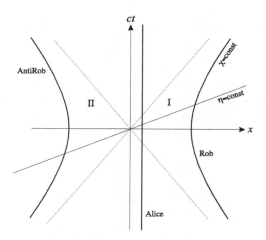

Fig. 1. Rindler space-time diagram: lines of constant position $\chi = $ const. are hyperbolae and all curves of constant η are straight lines that pass through the origin. An uniformly accelerated observer Rob travels along a hyperbola constrained to either region I or region II.

In Lecture 2 we found that metric in Rindler coordinates takes the form

$$ds^2 = e^{2a\chi}(d\eta^2 - d\chi^2)$$

The factor $e^{2a\chi}$ is known as the conformal factor. A part from this factor, the metric has the same form as metric in Rindler coordinates. Therefore, the Klein-Gordon equation takes the form

$$(\partial_\eta^2 - \partial_\chi^2)\phi = 0$$

The solutions are again plane waves

$$\bar{u}_k^I = \frac{1}{2\pi\omega}e^{i(k\chi-\omega\eta)}$$

$$\bar{u}_k^{*I} = \frac{1}{2\pi\omega}e^{-i(k\chi-\omega\eta)}$$

with $\omega = |k|$ and $-\infty < k < \infty$. These solutions only have support in region I and therefore, are not a complete set of solutions. The solutions $\bar{u}_k^I$ and $\bar{u}_k^{*I}$ are identified as positive and negative frequency solutions, respectively, with respect to the timelike Killing vector field ∂_η. The transformation which defines Rindler region II also gives rise to the spacetime $ds^2 = e^{2a\chi}(d\eta^2 - d\chi^2)$. However, in this case as t grows η becomes small. The timelike Killing vector field in this case is $-\partial_\eta$ and the solutions are

$$u_k^{II} = \frac{1}{\sqrt{2\pi\omega}}e^{i(k\chi+\omega\eta)}$$

$$u_k^{*II} = \frac{1}{\sqrt{2\pi\omega}}e^{-i(k\chi+\omega\eta)}$$

with support in region II. The solutions of region I together with the solutions in region II form a complete set of orthonormal solutions. Therefore we can quantize the field in this basis. We thus, obtain that the field in Rindler coordinates is given by

$$\hat{\phi} = \int (u_k^I a_k^I + u_k^{II} a_k^{II} + h.c.)dk$$

The vacuum state in the Rindler basis is $|0\rangle_R = |0\rangle_I \otimes |0\rangle_{II}$ where $a_k^I |0\rangle^I = 0$ and $a_k^{II} |0\rangle^{II} = 0$.

We would now like to express the Minkowski vacuum $|0\rangle^{\mathcal{M}}$ in terms of Rindler states. For this we make use of the inner product and find the Bogoliubov coefficents, obtaining

$$a_k = \int ((u_k, u_{k'}^I)a_k^I + (u_k, u_{k'}^{*I})a_k^{\dagger I} + (u_k, u_{k'}^{II})a_k^{II} + (u_k, u_{k'}^{*II})a_k^{\dagger II})dk$$

where, for example, $(u_k, u_k^I) = \int(u_k\partial_t u_{k'}^{I\dagger} - \partial_t u_k u_{k'}^{I\dagger})dx$. This calculation is quiet involved. The Minkowski creation and annihilation operators result in a infinite sum of Rindler operators. However, by introducing an alternative basis for the inertial observers, the Unruh basis, the problem is significantly simplified. The Unruh solutions correspond to

$$u_k^U = \cosh r u_k^I + \sinh r u_k^{II*},$$

where $\operatorname{sech}^2(r) = 1 - e^{\frac{2\pi\omega}{a}}$. An similar procedure to the one followed above must be carried out between Unruh solutions and Minkowski solutions. We then find that the Unruh annihilation operators result in an intergal of Minkowski annihilation operators of the form

$$A_k = \int C_{k'} a_{k'} dk'.$$

Therefore, the Unruh vacuum and the Minkowski vacuum coincide, i.e. $A_k |0\rangle^{\mathcal{M}} = 0$. The transformation between Unruh and Rindler operators yields

$$A_k = \cosh(r) a_k^I - \sinh(r) a_k^{II\dagger}$$

This transformation is much simpler since it involves a single Unruh and Rindler frequency. Considering the Ansatz

$$|0_k\rangle^{\mathcal{M}} = \sum_n A_n |n_k\rangle^I |n_k\rangle^{II}$$

we can find the Minkowski vacuum in terms of Rindler states by solving the equation

$$0 = A_k |0_k\rangle^{\mathcal{M}} = (\cosh(r) a_k^I - \sinh(r) a_k^{II\dagger}) \sum_n A_n |n_k\rangle^I |n_k\rangle^{II}$$

this yields

$$0 = \sum A_n \cosh(r) \sqrt{n} |(n-1)_k\rangle^I |n_k\rangle^{II}$$
$$- \sum A_n \sinh(r) \sqrt{n+1} |n_k\rangle^I |(n+1)_k\rangle^{II}$$
$$= \sum A_{n+1} \cosh(r) \sqrt{n+1} |n_k\rangle^I |(n+1)_k\rangle^{II}$$
$$- \sum A_n \sinh(r) \sqrt{n+1} |n_k\rangle^I |(n+1)_k\rangle^{II}$$

We obtain the following recurrence relation

$$A_{n+1} = \frac{\sinh(r)}{\cosh(r)} A_n \Rightarrow A_n = \tanh^n(r) A_0$$

and therefore,

$$|0_k\rangle^{\mathcal{M}} = \sum A_0 \tanh^n(r) |n_k\rangle^I |n_k\rangle^{II}$$

From the normalization condition for the vacuum state we obtain

$$|A_0|^2 \sum_n \tanh^{2n}(r) = 1.$$

Since

$$|A_0|^2 \sum_n \tanh^{2n}(r) = |A_0|^2 \frac{1}{1 - \tanh^2(r)}$$
$$= |A_0|^2 \cosh^2(r)$$

we finally find

$$|0_k\rangle^{\mathcal{M}} = \frac{1}{\cosh(r)} \sum_n \tanh^n(r) |n_k\rangle^I |n_k\rangle^{II}$$

The state in the Rindler basis corresponds to a two mode squeezed state. Here $\tanh r = e^{-\frac{2\pi\omega}{a}}$. Since the accelerated observer is constrained to move in region I we must trace over the states in region II. We will carry out this calculation in the next lecture. The result will be that in the Rindler wedge I the state is a thermal state. This is the well known Unruh effect: while inertial observers describe the state of the field to be the vacuum, observers in uniform acceleration observe a thermal state.

REMARK Therefore, the particle content of a field is observer dependent.

REFERENCES

- P. C. W. Davis, J. of Phys.,(A8): 609, 1975.
- W. G. Unruh, Phys. Rev. D, (1): 870, 1976.
- *Quantum Field Theory in Curved Spacetime* by N.D. Birrell and P.C.W. Davies, CUP (1982).

7. Entanglement in Flat Spacetime

In the last lecture we learned that while an inertial observer detects the field to be in the vacuum state, the state in the Rindler basis corresponds to a two mode squeezed state. In this lecture we will show that from the perspective of observers in uniform acceleration constrained to move in Rindler region I, the field is in a thermal state with temperature. We will analyze the effects of this on entanglement. For this, we will consider an entangled state between two modes of the field and show that while for inertial observers the state is maximally entangled, the entanglement degrades when considering observers in uniform acceleration. We found that the inertial vacuum state corresponds to a two mode squeezed state in the Rindler basis

$$|0_k\rangle^{\mathcal{M}} = \frac{1}{\cosh(r)} \sum_n \tanh^n(r) |n_k\rangle^I |n_k\rangle^{II}$$

The density matrix of this state is given by $\rho_0 = |0_k\rangle \langle 0_k|^{\mathcal{M}}$. Given that we are working with a single mode we will, for now, drop the frequency index k. Since an observer constrained to move in region I has no access to information in region II, he must trace over the states in region II. Therefore, the state in region I corresponds to the following reduced density matrix

$$\begin{aligned}
\rho_I &= Tr_{II}\left(\frac{1}{\cosh^2(r)} \sum_{n,m} \tanh^{n+m}(r) |n\rangle^I |n\rangle^{II} \langle m|^I \langle m|^{II}\right) \\
&= \frac{1}{\cosh^2(r)} \sum_{n,m} \tanh^{n+m}(r) \langle m|n\rangle^I |n\rangle^{II} \langle m|^{II} \\
&= \frac{1}{\cosh^2(r)} \sum_{n,m} \tanh^{n+m}(r) \delta_{mn} |n\rangle^{II} \langle m|^{II} \\
&= \frac{1}{\cosh^2(r)} \sum_n \tanh^{2n}(r) |n\rangle^{II} \langle n|^{II}.
\end{aligned}$$

Since $\tanh r = e^{-\frac{2\pi\omega}{a}}$ and $\operatorname{sech}^2(r) = 1 - e^{\frac{2\pi\omega}{a}}$, the density matrix corresponds to a canonical thermal state

$$\rho_I = (e^{-\frac{2\pi\omega}{a}} - 1) \sum_n (e^{-\frac{2\pi\omega}{a}})^n |n\rangle^{II} \langle n|^{II}.$$

with temperature $T_U = \frac{a}{2\pi k_B}$ (where k_B is the Boltzman constant) proportional to the observer's acceleration. The temperature is known as the Unruh temperature.

We now analyze the effects of this on the entanglement between two field modes of the field. In the canonical scenario considered in the study of entanglement in non-inertial frames the field, from the inertial perspective, is considered to be in a state where all modes are in the vacuum state except for two of them which are in a two-mode entangled state. For example, the Bell state

$$|\Psi\rangle^{\mathcal{U}} = \frac{1}{\sqrt{2}} \left(|0_k\rangle^{\mathcal{U}} |0_{k'}\rangle^{\mathcal{U}} + |1_k\rangle^{\mathcal{U}} |1_{k'}\rangle^{\mathcal{U}}\right), \tag{9}$$

where $\mathcal{U}$ labels Unruh states and k, k' are two Unruh frequencies. Two inertial observers, Alice and Bob, each carrying a monocromatic detector sensitive to frequencies k and k' respectively, would find maximal correlations in their measurements since the Bell state is maximally entangled. It is then interesting to investigate to what degree the state is entangled when described by observers in uniform acceleration. In the simplest scenario, Alice is again considered to be inertial and an uniformly accelerated observer Rob is introduced, who carries a monocromatic detector sensitive to mode k'. To study this situation, the states corresponding to Rob must be transformed into the appropriate basis, in this case, the Rindler basis. Note that, from the inertial perspective, we employ the Unruh basis, which is an alternative basis for inertial observers, since the transformation into the Rindler basis is very simple. We have already calculated the transformation for the vacuum state, and with that in hand, we can calculate the single particle state $|1_{k'}\rangle^{\mathcal{U}} = A_{k'}^{\dagger} |0_{k'}\rangle^{\mathcal{U}}$. This calculation will be left as an exercise. The resulting state must be

$$A_{k'}^{\dagger} |0_{k'}\rangle^{\mathcal{U}} = |1_{k'}\rangle^{\mathcal{U}} = \frac{1}{\cosh(r)} \sum_n \tanh^n(r)\sqrt{n+1}\,|(n+1)_{k'}\rangle_I\,|n_{k'}\rangle_{II}.$$

Thus, the maximally entangled state from the perspective of inertial Alice and accelerated Rob is

$$|\Psi\rangle^{\mathcal{U}} = \frac{1}{\sqrt{2}}\,|0_k\rangle \otimes \frac{1}{\cosh(r)} \sum_n \tanh^n(r)\,|n_{k'}\rangle^I\,|n_{k'}\rangle^{II}$$
$$+\frac{1}{\sqrt{2}}\,|1_k\rangle \otimes \frac{1}{\cosh(r)} \sum_n \tanh^n(r)\sqrt{n+1}\,|(n+1)_{k'}\rangle^I\,|n_{k'}\rangle^{II}.$$

Since Rob is causally disconnected from region II we must take the trace over region II. The density matrix for the Alice-Rob subsystem is

$$\rho_{AR} = \frac{1}{\cosh^2(r)} \sum_{n=0}^{\infty} \tanh^{2n}(r)\rho_n$$

where

$$\rho_n = |0_k, n_{k'}\rangle\,\langle 0_k, n_{k'}|$$
$$+\frac{\sqrt{n+1}}{\cosh(r)}\left(|0_k, n_{k'}\rangle\,\langle 1_k, (n+1)_{k'}| + |1_k, (n+1)_{k'}\rangle\,\langle 0_k, n_{k'}|\right)$$
$$+\frac{n+1}{\cosh^2(r)}\,|1_k, (n+1)_{k'}\rangle\,\langle 1_k, (n+1)_{k'}|$$

We calculate the entanglement between the modes detected by Alice and Rob using the logarithmic negativity

$$E_N(\rho_{AR}) = \log_2 \|\rho_{AR}^{PT}\|$$

where $||\rho_{AR}^{PT}||$ is the trace norm of the partial transpose of the density matrix for Alice and Rob. Recall that the trace norm is equivalent to summing over the negative eigenvalues of the matrix. Therefore, we must find the eigenvalues of the partial transpose of the Alice-Rob density matrix. The partial transpose is found by exchanging the states corresponding to Alice

$$\rho_{AR}^{PT} = \frac{1}{\cosh^2(r)} \sum_{n=0}^{\infty} \tanh^{2n}(r) \rho_n^{PT}$$

with

$$
\begin{aligned}
\rho_n^{PT} &= |0_k, n_{k'}\rangle \langle 0_k, n_{k'}| \\
&+ \frac{\sqrt{n+1}}{\cosh(r)} \left(|1_k, n_{k'}\rangle \langle 0_k, (n+1)_{k'}| + |0_k, (n+1)_{k'}\rangle \langle 1_k, n_{k'}| \right) \\
&+ \frac{n+1}{\cosh^2(r)} |1_k, (n+1)_{k'}\rangle \langle 1_k, (n+1)_{k'}|
\end{aligned}
$$

This matrix is infinite dimensional, however it has a block diagonal form which allows us to diagonalize the matrix block by block. Considering the density matrix for the $(n, n+1)$ sector, each block takes the form

$$
\rho_n^{PT} = \frac{1}{\cosh^2(r)}
$$
$$
\times \begin{pmatrix}
\tanh^{2n}(r) & 0 & 0 & 0 \\
0 & \frac{n}{\cosh(r)} \tanh^{2(n-1)}(r) & \frac{\sqrt{n+1}}{\cosh(r)} \tanh^{2n}(r) & 0 \\
0 & \frac{\sqrt{n+1}}{\cosh(r)} \tanh^{2n}(r) & \tanh^{2(n+1)}(r) & 0 \\
0 & 0 & 0 & \frac{n+1}{\cosh^2(r)} \tanh^{2n}(r)
\end{pmatrix}
$$

were we have used the basis $\{|0_k, n_{k'}\rangle, |0_k, (n+1)_{k'}\rangle, |1_k, n_{k'}\rangle, |1_k, (n+1)_{k'}\rangle\}$. We are looking for the negative eigenvalues of this matrix. We can see that the eigenvalues corresponding to the first and last diagonal entries of the matrix are always positive. Therefore, we must simply diagonalize the 2x2 matrix

$$
\frac{1}{\cosh^2(r)}
\begin{pmatrix}
\frac{n}{\cosh(r)} \tanh^{2(n-1)}(r) & \frac{\sqrt{n+1}}{\cosh(r)} \tanh^{2n}(r) \\
\frac{\sqrt{n+1}}{\cosh(r)} \tanh^{2n}(r) & \tanh^{2(n+1)}(r).
\end{pmatrix}
$$

The eigenvalues of the density matrix are

$$
\lambda_\pm = \frac{\tanh^n(r)}{4\cosh^2(r)} \left(\frac{n}{\sinh^2(r)} + \tanh^2(r) \pm Z_n \right)
$$

with

$$Z_n = \left(\frac{n}{\sinh^2(r)} + \tanh^2(r) \right)^2 + \frac{4}{\cosh^2(r)}.$$

Since one of the eigenvalues is always negative for finite r, the state is always entangled. To calculate the logarithmic negativity we sum over all negative eigenvalues and find $\mathcal{N}(\rho_{AR}) = \log_2((1/2\cosh^2(r)) + \Sigma)$ where

$$\Sigma = \sum_n \frac{\tanh^{2n}(r)}{2\cosh^2(r)} \sqrt{\left(\frac{n}{\sinh^2(r)} + \tanh^2(r) \right)^2 + \frac{4}{\cosh^2(r)}}$$

We conclude that entanglement is degraded when one of the observers moves in uniform acceleration. This means that entanglement is observer dependent. In the flat case, one can conclude that this is an effect of Rob's acceleration. Rob must be in a spaceship to be accelerated and energy must be supplied into the system. One can argue that, in the flat case, inertial observers play a special role and that therefore, a well defined notion of entanglement corresponds to the entanglement described from the inertial perspective. However, in curved spacetime different inertial observers describe a different particle content in the field which results in different degrees of entanglement in the field. In that case, there is no well-defined notion of entanglement.

REFERENCES

I. Fuentes-Schuller and R.B. Mann, Phys. Rev. Lett., (95): 120404, 2005.

8. Particle Creation in an Expanding Universe

In this lecture we will consider an example of a curved spacetime which does not admit a global timelike Killing vector field. However, the spacetime has two assymthoticaly flat regions in which timelike Killing vector fields can be found and therefore positive and negative solutions to the Klein-Gordon equations can be distinguished. We will consider a $(1 + 1)$-dim expanding Robertson-Walker universe which is asymptotically flat in the future and past infinity. Since in the past infinity spacetime is flat, particles states can be defined. In this region, we will consider the field to be in the vacuum state. We will then analyze the state from the perspective of observers in the future infinity. We will show that in the future infinity there has been particle creation.

The spacetime of a Robertson-Walker Universe in $(1 + 1)$-dim is given by

$$ds^2 = dt^2 - a^2(t)d\chi^2$$

where the spatial sections of the space time are expanding (or contracting) uniformly according to the function $a^2(t)$. Considering the infintesimal coordinate transformation

$$d\eta = \frac{dt}{a(t)}$$

the metric is written as $ds^2 = a^2(t)(d\eta^2 - d\chi^2)$. Defining $a^2(t) = c^2(\eta)$ we obtain the metric

$$ds^2 = c^2(\eta)(d\eta^2 - d\chi^2)$$

We now suppose that $c(\eta) = 1 + \epsilon(1 + \tanh(\sigma\eta))$ where ϵ and σ are constants. This describes a toy model of a universe undergoing a period of smooth expansion. The parameter ϵ is known as the expansion volume and σ is the expansion rate. In the limit $\eta \to -\infty$ the metric is $ds^2 = (d\eta^2 - d\chi^2)$ and in the limit $\eta \to -\infty$ then $ds^2 = (1 + 2\epsilon)(d\eta^2 - d\chi^2)$. Therefore, the metric is flat in these regions and the vector field ∂_η has Killing properties in these regions.

We now consider the massive Klein-Gordon equation $(\Box + m)\phi = 0$ in the Robertson-Walker spacetime described above. Here m is the mass of the field. Since the d'Alambertian in a curved spacetime is defined by

$$\Box\phi := \frac{1}{\sqrt{-g}}\partial_\mu(\sqrt{-g}g^{\mu\nu}\partial_\nu\phi)$$

in this spacetime the metric tensor g_{ab} has components

$$g_{ab} = c(\eta)\begin{pmatrix} 1 & 0 \\ 0 & -1 \end{pmatrix}$$

therefore $g = det(g_{ab}) = -c^2(\eta)$ and the contravariant metric is given by

$$g^{ab} = \frac{1}{c(\eta)}\begin{pmatrix} 1 & 0 \\ 0 & -1 \end{pmatrix}.$$

Hence

$$\Box\phi = \frac{1}{c(\eta)}\partial_\mu(c(\eta)g^{\mu\nu}\partial_\nu\phi)$$

$$= \frac{1}{c(\eta)}[\partial_0 c(\eta)g^{00}\partial_0 + \partial_1 c(\eta)g^{11}\partial_1]\phi$$

$$= \frac{1}{c(\eta)}[\partial_\eta c(\eta)\frac{1}{c(\eta)}\partial_\eta + \partial_\chi c(\eta)\frac{1}{c(\eta)}\partial_\chi]\phi$$

$$= \frac{1}{c(\eta)}[\partial_\eta^2 - \partial_\chi^2]\phi$$

Therefore, the Klein-Gordon equation takes the form

$$((\partial_\eta^2 - \partial_\chi^2) + c(\eta)m)\phi = 0$$

Exploiting the resulting spacial translational invariance we separate the solutions into

$$u_k = \frac{1}{\sqrt{2\pi\omega}}e^{ik\chi}\xi_k(\eta).$$

The equation then becomes

$$(\frac{1}{\sqrt{2\pi\omega}}e^{ik\chi}\partial_\eta^2\xi_k(\eta) + \frac{k^2}{\sqrt{2\pi\omega}}e^{ik\chi}\xi_k(\eta) + c(\eta)m\frac{1}{\sqrt{2\pi\omega}}e^{ik\chi}\xi_k(\eta)) = 0$$

and therefore,

$$\partial_\eta^2\xi_k(\eta) + (k^2 + c(\eta)m^2)\xi_k(\eta) = 0.$$

This equation can be solved for the whole spacetime in terms of two hypergeometric functions. We find two types of solutions

$$u_k^{(1)}(\eta,\chi) = \frac{1}{\sqrt{4\pi\omega_{in}}}e^{ik\chi - i\omega_+\eta\left(\frac{i\omega_-}{\sigma}\right)\ln 2\cosh(\sigma\eta)}{}_2F_2(\alpha,\beta,\gamma_1,\delta)$$

$$u_k^{(2)}(\eta,\chi) = \frac{1}{\sqrt{4\pi\omega_{out}}}e^{ik\chi - i\omega_+\eta\left(\frac{i\omega_-}{\sigma}\right)\ln 2\cosh(\sigma\eta)}{}_2F_1(\alpha,\beta,\gamma_2,\delta)$$

where the constants above are defined as

$$\alpha = 1 + \frac{i\omega_-}{\rho}, \quad \beta = \frac{i\omega_-}{\rho}, \quad \gamma_1 = 1 - \frac{i\omega_{in}}{\rho}, \quad \gamma_2 = 1 + \frac{i\omega_{out}}{\rho}$$

$$\delta = \frac{1}{2}(1 - \tanh(\rho\eta))$$

and the frequencies

$$\omega_{in} = [k^2 + m^2]^{\frac{1}{2}}$$

$$\omega_{out} = [k^2 + m^2(1 + 2\epsilon)]^{\frac{1}{2}}$$

$$\omega_\pm = \frac{1}{2}(\omega_{out} \pm \omega_{in}).$$

We note that in the limit $\eta \to -\infty$ the first solution becomes

$$u_k^{(1)} \to \frac{1}{\sqrt{4\pi\omega_{in}}} e^{ik\chi - i\omega_{in}\eta}$$

And in the case $\eta \to +\infty$ the second solution is

$$u_k^{(2)} \to \frac{1}{\sqrt{4\pi\omega_{out}}} e^{ik\chi - i\omega_{out}\eta}$$

One can see that in these limits the asymptotic solutions $u_k^{(1)}$ and $u_k^{(2)}$ to the Klein-Gordon equation are plane waves which can then be associated with positive mode solutions. The negative mode solutions correspond to u_k^{1*} and u_k^{2*}. Since u_k^1 is associated with a plan wave at $\eta \to -\infty$ (past infinity) we call these solutions in-waves $u_k^{(1)} \equiv u_k^{(in)}$. The solutions $u_k^{(2)} \equiv u_k^{(out)}$ which are associated with plane waves at $\eta \to +\infty$ (future infinity) will be called out-waves.

Using the linear transformation properties of hypergeometric functions we can write $u_k^{(in)}$ in terms of $u_k^{(out)}$. This is easier than caluclating the Bogolubov coefficients using the inner product. We then obtain

$$u_k^{(in)}(\eta, \chi) = \alpha_k u_k^{(out)}(\eta, \chi) + \beta_k u_k^{(out*)}(\eta, \chi)$$

where

$$\alpha_k = \left(\frac{\omega_{out}}{\omega_{in}}\right)^{\frac{1}{2}} \frac{\Gamma\left(1 - \frac{i\omega_{in}}{\sigma}\right) \Gamma\left(-\frac{i\omega_{out}}{\sigma}\right)}{\Gamma\left(-\frac{i\omega_+}{\sigma}\right) \Gamma\left(1 - \frac{i\omega_+}{\sigma}\right)}$$

$$\beta_k = \left(\frac{\omega_{out}}{\omega_{in}}\right)^{\frac{1}{2}} \frac{\Gamma\left(1 - \frac{i\omega_{in}}{\sigma}\right) \Gamma\left(-\frac{i\omega_{out}}{\sigma}\right)}{\Gamma\left(-\frac{i\omega_-}{\sigma}\right) \Gamma\left(1 - \frac{i\omega_-}{\sigma}\right)}$$

Here Γ are Gamma functions. From the above expressions we can read off the Bogolubov coefficients $\alpha_{kk'} = \alpha_k \delta_{kk'}$ and $\beta_{kk'} = \beta_k \delta_{-kk'}$. Therefore the transformation between annihilation operators yields

$$a_k^{in} = \alpha_k^* a_k^{out} - \beta_k^* a_{-k}^{out\dagger}.$$

We now consider that the state of the field in the past infinity is the vacuum state (no entanglement)

$$|0\rangle^{in} = \bigotimes_{k=-\infty}^{\infty} |0\rangle_k^{in}$$

and use the expression for the in-mode annihilation operator to calculate the state in the future infinity. Since the transformation between in and out annihilation operators only mixes modes of frequency k and $-k$, we consider the following Ansatz for the state in the future infinity

$$|0\rangle^{in} = \sum_n A_n |n\rangle_k^{out} |n\rangle_{-k}^{out}$$

Given that the vacuum state is defined as $a_k^{in} |0\rangle^{in} = 0$ we obtain the equation

$$a_k^{in} |0\rangle^{in} = \alpha_k^* \sum_n A_n \sqrt{n} |n-1\rangle_k^{out} |n\rangle_{-k}^{out} - \beta_k^* \sum_n A_n \sqrt{n} |n\rangle_k^{out} |n-1\rangle_{-k}^{out}$$

$$= \alpha_k^* \sum_n A_{n+1} \sqrt{n+1} |n\rangle_k^{out} |n+1\rangle_{-k}^{out}$$

$$- \beta_k^* \sum_n A_{n+1} \sqrt{n+1} |n+1\rangle_k^{out} |n\rangle_{-k}^{out}.$$

We then obtain the following recurrence relation

$$A_{n+1} = \frac{\beta_k^*}{\alpha_k^*} A_n \implies A_n = \left(\frac{\beta_k^*}{\alpha_k^*}\right)^n A_0.$$

The vacuum state from the perspective of observers in the past infinity becomes

$$|0\rangle^{in} = A_0 \sum_n \left(\frac{\beta_k^*}{\alpha_k^*}\right)^n |n\rangle_k^{out} |n\rangle_{-k}^{out}$$

for observers in the future infinity. Employing the normalization condition $\langle 0|0\rangle^{in} = 1$ and defing $\gamma = \left|\frac{\beta_k}{\alpha_k}\right|^2$ we obtain

$$|A_0|^2 = 1 - \gamma$$

and therefore,

$$|0\rangle^{in} = \sqrt{1-\gamma} \sum_n \gamma^n |n\rangle_k^{out} |n\rangle_{-k}^{out}.$$

In terms of the explicit Bobolubov coefficients

$$\gamma = \frac{\sinh^2(\pi\omega_-/\sigma)}{\sinh^2(\pi\omega_+/\sigma)}.$$

REMARKS The vacuum state from the perspective of observers in the remote past has particles in the remote future. Due to the expansion of the universe there has been particle creation.

REFERENCES

Quantum Field Theory in Curved Spacetime by N.D. Birrell and P.C.W. Davies, CUP (1982).

9. Entanglement in Curved Spacetime

In the last lecture we considered an example of curved spacetime where particles can be defined in two asymptotically flat regions regions. We found that while in the remote past the field was in the vacuum state, in the remote future particles have been created. We recall however, that in the interim region, where the universe is undergoing expansion, no sensible notion of particles exist. In this lecture we will continue with this example and show that in the future infinity entanglement has been created between field modes. Interestingly, it is possible to learn about the expansion parameters of the Universe from the entanglement generated. As a last example, we will consider the entanglement between two field modes in the spacetime of an eternal black hole. We will see that while two observers falling into the black hole describe the state of the field to be in a maximally entangled state, the entanglement in the state becomes degraded from the perspective of an inertial observer, Alice, who falls into a black hole and an non-inertial observer, Rob, escaping the black hole.

9.1. *Entanglement in an expanding universe*

We found that the state corresponding to the vacuum state in the past infinity corresponds to the following two mode state in the future infinity

$$|0\rangle^{in} = \sqrt{1-\gamma} \sum_n \gamma^n |n\rangle_k^{out} |n\rangle_{-k}^{out}.$$

Since the state is pure, we can employ the Von-Neuman entropy to quantify the entanglement generated the field modes k and $-k$. In order to do this we need to compute the reduced density matrix for one of the modes. The densitity matrix for the state is

$$\rho_0 = |0\rangle^{in} \langle 0|^{in}$$
$$= (1-\gamma) \sum_{n,m} \gamma^{(n+m)} |n\rangle_k^{out} |n\rangle_{-k}^{out} \langle m|_k^{out} \langle m|_{-k}^{out}.$$

The reduced density matrix for mode k is obtained by tracing over mode $-k$

$$\rho_k = tr_{-k}[\rho_0]$$
$$= (1-\gamma) \sum_{n,m} \gamma^{(n+m)} \delta_{nm} |n\rangle_k^{out} \langle m|_k^{out}$$
$$= (1-\gamma) \sum_{n,m} \gamma^{2n} |n\rangle_k^{out} \langle n|_k^{out}$$

Since the reduced denstiy matrix is already in diagonal form with eigenvalues $\lambda_n = (1 - \gamma)\gamma^{2n}$, it is straight forward to compute the Von-Neumann entropy

$$
\begin{aligned}
S(\rho_k) &= -Tr\left(\rho_k \log_2 \rho_k\right) \\
&= -(1 - \gamma) \sum_n \gamma^{2n} \log_2((1 - \gamma)\gamma^{2n}) \\
&= (\gamma - 1) \sum_n (2n\gamma^{2n} \log_2 \gamma + \gamma^{2n} \log_2(1 - \gamma)) \\
&= 2(\gamma - 1) \log_2 \gamma \sum_n n\gamma^{2n} + (\gamma - 1) \log_2(1 - \gamma) \sum_n \gamma^{2n} \\
&= (\gamma - 1) \log_2 \gamma \sum_n 2n\gamma^{2n} + (\gamma - 1) \log_2(1 - \gamma)\frac{1}{1 - \gamma} \\
&= (\gamma - 1) \log_2 \gamma \frac{\partial}{\partial \gamma}(1 - \gamma)^{-1} - \log_2(1 - \gamma)
\end{aligned}
$$

which yields

$$
S(\rho_k) = \log_2\left[\frac{\gamma^{\frac{\gamma}{\gamma-1}}}{1 - \gamma}\right]
$$

Entanglement has been created in the remote future due to the expansion of the universe. The entanglement depends on the cosmological constants since the coefficient γ depends on the expansion rate σ, the expansion volume ϵ and the frequency of the modes involved through

$$
\gamma = \frac{\sinh^2(\pi\omega_-/\sigma)}{\sinh^2(\pi\omega_+/\sigma)}
$$

with

$$
\begin{aligned}
\omega_\pm &= \frac{1}{2}(\omega_{out} \pm \omega_{in}) \\
\omega_{in} &= [k^2 + m^2]^{\frac{1}{2}} \\
\omega_{out} &= [k^2 + m^2(1 + 2\epsilon)]^{\frac{1}{2}}.
\end{aligned}
$$

In the case of light particles, the equations can be inverted and we can show that we can estimate the expansion parameters from the entanglement.

REMARK Entanglement has been created between field modes due to the expansion of the universe. It is possible to learn from the past history of the universe from the entanglement in the modes.

9.2. *Alice falls into a black hole*

The Schwarzchild spacetime of an eternal black hole describes the geometry of a spherical non-rotating mass m. Considering only the radial component, the metric is

$$ds^2 = (1 - (2m/R))dT^2 - (1 - (2m/R))^{-1}dR^2.$$

The spacetime is curved and admits no global timelike Killing vector fields. However, close to the horizon of the black hole defined by $R = 2m$, the space time is approximately flat. To see this, we consider the following coordinate change $R - 2m = x^2/8m$, such that

$$1 - (2m/R) = x^2/8mR = \frac{(x^2/8m)}{(x^2/8m + 2m)} = \frac{(Ax)^2}{(1 + (Ax)^2)} \approx (Ax)^2$$

when $x \approx 0$ with $A = 1/4m$. This means that $dR^2 = (Ax^2)dx^2$. Therefore, very close to the horizon $R \approx 2m$ the Schwarzchild spacetime can be approximated by Rindler space

$$ds^2 = -(Ax)^2 dT^2 + dx^2.$$

where the acceleration parameter $a = A^{-1}$. This means that, very close to the horizon of the black hole, we can consider Alice being inertial and falling into the black hole while Rob escapes the fall by being accelerated. If Alice claims that the state of the field is the vacuum state, then Rob detects a thermal state since the state has the form

$$|0_k\rangle^{\mathcal{M}} = \frac{1}{\cosh(r)} \sum_n \tanh^n(r) \, |n_k\rangle^{in} \, |n_k\rangle^{out}.$$

were r is a function of the mass of the black hole. Here *in* and *out* denote the modes inside and outside the black hole. $|0_k\rangle^{\mathcal{M}}$ is the state detected by Alice. If we then consider that the field, from Alice's perspective is a maximally entangled state of two modes, Rob will detect less entanglement between the modes due to the Hawking effect.

10. Final Remarks

The phenomenon of entanglement has been extensively studied in non-relativistic settings. Much of the interest on this quantum property has stemmed from its relevance in quantum information theory. However, relatively little is known about relativistic effects on entanglement despite the fact that many of the systems used in the implementation of quantum information involve relativistic systems such as photons. The vast majority of investigations on entanglement assume that the world is flat and non-relativistic. Understanding entanglement in spacetime is ultimately necessary because the world is fundamentally relativistic. Moreover, entanglement plays a prominent role in black hole thermodynamics and in the information loss problem.

The question of understanding entanglement in non-inertial frames has been central to the development of the emerging field of relativistic quantum information. The main aim of this field is to incorporate relativistic effects to improve quantum information tasks (such as quantum teleportation) and to understand how such protocols would take place in curved space-times. In most quantum information protocols entanglement plays a prominent role. Therefore, it is of a great interest to understand how it can be degraded or created by the presence of horizons or spacetime dynamics.

In this series of lectures we showed that studies on relativistic entanglement show that conceptually important qualitative differences to a non-relativistic treatment arise. For instance, entanglement was found to be an observer-dependent property that is degraded from the perspective of accelerated observers moving in flat spacetime. These results suggest that entanglement in curved spacetime might not be an invariant concept.

REFERENCES

J. Ball, I. Fuentes-Schuller, and F. P. Schuller, Phys. Lett. A, (359): 550, 2006.

11. Problems

11.1. *Entanglement of Dirac fields in non-inertial frames*

In a parallel analysis to the bosonic case, we consider a Dirac field ϕ satisfying the equation $\{i\gamma^\mu(\partial_\mu - \Gamma_\mu) + m\}\phi = 0$ where γ^μ are the Dirac-Pauli matrices and Γ_μ are spinorial affine connections. The field expansion in terms of the Minkowski solutions of the Dirac equation is

$$\phi = N_M \int \left(c_{k,M} \, u^+_{k,M} + d^\dagger_{k,M} \, u^-_{k,M} \right) dk, \tag{10}$$

Where N_M is a normalisation constant and the label $\pm$ denotes respectively positive and negative energy solutions (particles/antiparticles) with respect to the Minkowskian Killing vector field ∂_t. The label k is a multilabel including energy and spin $k = \{E_\omega, s\}$ where s is the component of the spin on the quantisation direction. c_k and d_k are the particle/antiparticle operators that satisfy the usual anticommutation rule

$$\{c_{k,\mathrm{M}}, c^\dagger_{k',\mathrm{M}}\} = \{d_{k,\mathrm{M}}, d^\dagger_{k',\mathrm{M}}\} = \delta_{kk'}, \tag{11}$$

and all other anticommutators vanishing.

The Dirac field operator in terms of Rindler modes is given by

$$\phi = N_{\mathrm{R}} \int \left(c_{j,\mathrm{I}} u^+_{j,\mathrm{I}} + d^\dagger_{j,\mathrm{I}} u^-_{j,\mathrm{I}} + c_{j,\mathrm{II}} u^+_{j,\mathrm{II}} + d^\dagger_{j,\mathrm{II}} u^-_{j,\mathrm{II}}\right) dj, \tag{12}$$

Where N_{R} is, again, a normalisation constant. $c_{j,\Sigma}, d_{j,\Sigma}$ with $\Sigma = \mathrm{I}, \mathrm{II}$ represent Rindler particle/antiparticle operators. The usual anticommutation rules again apply. Note that operators in different regions $\Sigma = \mathrm{I}, \mathrm{II}$ do not commute but anticommute. $j = \{E_\Omega, s'\}$ is again a multi-label including all the degrees of freedom. Here $u^\pm_{k,\mathrm{I}}$ and $u^\pm_{k,\mathrm{II}}$ are the positive/negative frequency solutions of the Dirac equation in Rindler coordinates with respect to the Rindler timelike Killing vector field in region I and II, respectively. The modes $u^\pm_{k,\mathrm{I}}$, $u^\pm_{k,\mathrm{II}}$ do not have support outside the right, left Rindler wedge. The annihilation operators $c_{k,\mathrm{M}}, d_{k,\mathrm{M}}$ define the Minkowski vacuum $|0\rangle_{\mathrm{M}}$ which must satisfy

$$c_{k,\mathrm{M}} |0\rangle_{\mathrm{M}} = d_{k,\mathrm{M}} |0\rangle_{\mathrm{M}} = 0, \qquad \forall k. \tag{13}$$

In the same fashion $c_{j,\Sigma}, d_{j,\Sigma}$, define the Rindler vacua in regions $\Sigma = \mathrm{I}, \mathrm{II}$

$$c_{j,\mathrm{R}} |0\rangle_\Sigma = d_{j,\mathrm{R}} |0\rangle_\Sigma = 0, \qquad \forall j, \ \Sigma = \mathrm{I}, \mathrm{II}. \tag{14}$$

As in the bosonic case, we will work with Unruh modes for the inertial observers, where the transformation between Unruh and Rindler operators is given by and the operators

$$C_k \equiv \left(\cos r_k \, c_{k,\mathrm{I}} - \sin r_k \, d^\dagger_{k,\mathrm{II}}\right).$$

It can be shown that for a massless Dirac field the Unruh operators have the same form as Eq. (15) however in this case $\tan r_k = e^{-\pi\Omega_a/a}$. In this case, to find the Minkowski vacuum in the Rindler basis we consider the following ansatz

$$|0\rangle_{\mathrm{M}} = \bigotimes_\Omega |0_\Omega\rangle_{\mathrm{M}}, \tag{15}$$

were

$$|0_\Omega\rangle_M = \sum_{n,s} \left(F_{n,\Omega,s} |n_{\Omega,s}\rangle^+_\mathrm{I} |n_{\Omega,-s}\rangle^-_\mathrm{II}\right)$$

$$\tag{16}$$

where the label $\pm$ denotes particle/antiparticle modes and s labels the spin. The minus signs on the spin label in region II show explicitly that spin, as all the magnitudes which change under time reversal, is opposite in region I with respect to region II.

Due to the anticommutation relations we must introduce the following sign conventions

$$
\begin{aligned}
|1_\Omega\rangle_{\mathrm{I}}^+ |1_\Omega\rangle_{\mathrm{II}}^- &= d_{\Omega,\mathrm{IV}}^\dagger c_{\Omega,\mathrm{I}}^\dagger |0_\Omega\rangle_{\mathrm{I}}^+ |0_\Omega\rangle_{\mathrm{II}}^-, \\
&= -c_{\Omega,\mathrm{I}}^\dagger d_{\Omega,\mathrm{II}}^\dagger |0_\Omega\rangle_{\mathrm{I}}^+ |0_\Omega\rangle_{\mathrm{II}}^-, \\
|1_\Omega\rangle_{\mathrm{I}}^- |1_\Omega\rangle_{\mathrm{II}}^+ &= c_{\Omega,\mathrm{IV}}^\dagger d_{\Omega,\mathrm{I}}^\dagger |0_\Omega\rangle_{\mathrm{I}}^- |0_\Omega\rangle_{\mathrm{II}}^+, \\
&= -d_{\Omega,\mathrm{I}}^\dagger c_{\Omega,\mathrm{II}}^\dagger |0_\Omega\rangle_{\mathrm{I}}^- |0_\Omega\rangle_{\mathrm{II}}^+.
\end{aligned}
\tag{17}
$$

We will now consider the simplest case that preserves the fundamental Dirac characteristics which corresponds to Grassman scalars. In this case the Pauli exclusion principle limits the sums to $n = 0, 1$ and there is no spin. For convenience, is it suitable to introduce the following notation,

$$
|nn'n''n'''\rangle_\Omega \equiv |n_\Omega\rangle_{\mathrm{I}}^+ |n_\Omega'\rangle_{\mathrm{II}}^- |n_\Omega''\rangle_{\mathrm{I}}^- |n_\Omega'''\rangle_{\mathrm{II}}^+.
\tag{18}
$$

- Considering Grassman scalars, obtain the form of the coefficients $F_{n,\Omega}$ for the vacuum by imposing that the Minkowski vacuum is annihilated by the particle annihilator C_k for all frequencies and values for the spin third component, ie. $C_\Omega |0_\Omega\rangle_{\mathrm{R}} = 0$.
- Obtain the Unruh one particle state by applying the creation operator to the vacuum state $|1_j\rangle_{\mathrm{U}} = C_\Omega^\dagger |0_j\rangle_{\mathrm{M}}$.
- Consider the following fermionic maximally entangled state

$$
|\Psi\rangle = \frac{1}{\sqrt{2}} \left(|0_\omega\rangle_{\mathrm{M}} |0_\Omega\rangle_{\mathrm{U}} + |1_\omega\rangle_{\mathrm{M}}^+ |1_\Omega\rangle_{\mathrm{U}}^+ \right),
\tag{19}
$$

which is the fermionic analog to the bosonic maximally entangled state studied in our lectures. Compute Alice-Rob partial density matrix by tracing over region II .

- Obtain the partial transpose of density matrix and find its negative eigenvalues. Note that the matrix is block diagonal and only two blocks contribute.
- Compute the negativity.
- Compare results with the bosonic case.

Acknowledgments

I would like to thank Prof. Mikio Nakahara and the organizers of the Summer School on Diversities in Quantum Computation/Information for the opportunity of presenting this lecture course. I thank David Bruschi, Eduardo Martín-Martínez, Andrzej Dragan and specially, Antony Lee and Frederic Schuller for their assistance in preparing these notes. Finally, I thank in advance all the participants attending the lectures for your attention.

References

1. A. M. Turing, Proc. Lond. Math. Soc. 2, **42**, 230 (1936)
2. M. Czachor, Phys. Rev. A **55**, 72 (1997).
3. following Czachor's work on relativistic EPR: D. Ahn, H. J. Lee, Y. H. Moon, and S. W. Hwang, Phys. Rev. A **67**, 012103 (2003); H. Terashima and M. Ueda, Int. J. Quant. Info. **1**, 93 (2003); S. Massar and P. Spindel, Phys. Rev. D **74**, 085031 (2006); J. Ball quant-ph/0601082 (2006); V. Pankovic. UNS-17/05, quant-ph/0506129 (2005).
4. A. Peres, P. F. Scudo and D. R. Terno, Phys. Rev. Lett. **88**, 230402 (2002); A. Peres and D. Terno, Rev. Mod. Phys. **76**, 93 (2004).
5. R. M. Gingrich and C. Adami, Phys. Rev. Lett. **89**, 270402 (2002); A. J. Bergou, R. M. Gingrich, and C. Adami, Phys. Rev. A **68**, 042102 (2003); J. Pachos and E. Solano, Quant. Inf. Comp. **3** 115 (2003).
6. I. Fuentes-Schuller and R. B. Mann, Phys. Rev. Lett. **95**, 120404 (2005).
7. P. M. Alsing, I. Fuentes-Schuller, R. B. Mann, and T. E. Tessier, Phys. Rev. A **74**, 032326 (2006).
8. G. Adesso, I. Fuentes-Schuller, and M. Ericsson, Phys. Rev. A. **76**, 062112 (2007).
9. P. M. Alsing and G. J. Milburn, Phys. Rev. Lett. **91**, 180404 (2003); P. M. Alsing and G. J. Milburn, Quant. Inf. Comp. **2**, 487 (2002); M. Czachor and M. Wilczewski, Phys. Rev. A **68**, 010302 (2003).
10. M. Han, S. Jay Olson, and J. P. Dowling, Phys. Rev. A **78** 022302 (2007).
11. M. Czachor and M. Wilczewski, Phys. Rev. A **68** 010302 (2003).
12. J. Barrett, L. Hardy, and A. Kent, Phys. Rev. Lett. **95**, 010503 (2005).
13. J. Ball, I. Fuentes-Schuller, and F. P. Schuller, Phys. Lett. A **359**, 550 (2006).
14. G. L. Ver Steeg1 and N. C. Menicucci, quant-ph 0711.3066 (2008).
15. Y. Shi, Phys. Rev. D **70**, 105001 (2004).
16. P. Kok, U. Yurtsever, S. L. Braunstein, and J. P. Dowling quant-ph/0206082 (2002).
17. G. Adesso and I. Fuentes-Schuller, Quant. Inf. Comput. **9**, 0657-0665 (2009); J. Preskill and D. Gottesman JHEP **0403**, 026 (2004); P. Hyden and J. Preskill, JHEP **0709**, 120 (2007); S. Lloyd, Phys. Rev. Lett. **96**, 061302 (2006); D. Ahn, JHEP **0703**, 021 (2007).
18. D. Terno J. Phys. Conf. Ser. **33**, 469-474 (2006); E. Livine and D. Terno, Rev. D **75** 084001 (2007); T. Konopka and F. Markopoulou gr-qc/0601028 (2006).

SYSTEMATIC CONSTRUCTION OF BELL-LIKE INEQUALITIES AND PROPOSAL OF A NEW TYPE OF TEST

SHOGO TANIMURA

Department of Applied Mathematics and Physics, Graduate School of Informatics
*Kyoto University, Kyoto 606-8501, Japan**

Although it is usually argued that the violation of the Bell inequality is a manifestation of the nonlocality of quantum state, we show that it is a manifestation of the noncommutativity of quantum observables that are defined at the same location. From this point of view we invent a method for systematic construction of generalized Bell inequalities and derive a new inequality that belongs to a different type from the traditional Bell inequality. The new inequality provides a severer and fairer test of quantum mechanics.[†]

Keywords: Hidden-variable theory; Bell inequality, Noncommutativity.

1. Introduction

Quantum mechanics is quite successful for explaining and predicting microscopic phenomena and its validity has been confirmed repeatedly. Nevertheless it contains conceptual difficulties like indeterminism and nonlocality, which have been annoying physicists.

Bohm[2] and other people[3] attempted to build an alternative theory of quantum mechanics, which is called the hidden-variable theory. In this theory, it is assumed that a physical system has a hidden variable. Once the value of the hidden variable is fixed, a value of any physical quantity is determined uniquely. However, the value of hidden variable can vary in each run of experiments and experimentalists do not know nor cannot control it. So, the outputs of measurements look random and physicists can calculate

---

*After 2011 April, move to Department of Complex Systems Science, Graduate School of Information Science, Nagoya University, Nagoya 464-8601, Japan. E-mail: tanimura@is.nagoya-u.ac.jp

[†]The full paper version of this work has been published in the journal.[1] But some corrections have been made in this lecture note.

only probabilities of various outputs. This is the basic idea of the hidden-variable theory. In a sense, the hidden-variable theory is an application of the ordinary classical probability theory in which the hidden variable is considered as a random variable.

The hidden-variable theory is built on the following assumptions: (1) The state of a physical system is completely specified not by wave function or state vector ψ, but by a variable λ or a set of variables $\lambda = (\lambda_1, \lambda_2, \cdots, \lambda_n)$. (2) Any physical observable A is a function of λ. Once the value of the hidden variable λ are specified, the value $A(\lambda) \in \mathbb{R}$ of the observable is uniquely determined. (3) A precise value of the variable λ in each experiment is not known. The variable λ obeys a probability distribution $P(\lambda)$, which is nonnegative, $P \geq 0$, and normalized, $\int P d\lambda = 1$. (4) Additionally, in the *local* hidden-variable theory, the value physical quantity cannot be affected by superluminous signals.

In this scheme, the expectation value of the observable A is calculated as

$$\langle A \rangle = \int A(\lambda) P(\lambda) \, d\lambda \tag{1}$$

with the probability distribution $P(\lambda)$ of the hidden variable λ. Thus, a question arises: *can we distinguish the hidden-variable theory from quantum mechanics by a physical experiment?* At the first glance, there are huge degrees of freedom in the choice of probability distribution of the hidden variables. Hence, it seems always possible to adjust the probability distribution of the hidden variable for reproducing the same prediction as the quantum-mechanical prediction. There seems no experimental test that can judge any discrepancy between quantum mechanics and the hidden-variable theory.

Bell[4,5] proved an inequality that bounds the expectation value of a certain observable in the scheme of the local hidden-variable theory. Clauser, Horne, Shimony and Holt[6,7] reformulated Bell's inequality in a form more suitable for experimental tests. They introduced four observables A_1, A_2, B_1, B_2, which take ± 1 as their values, and combined them to define a quantity $S := A_1 B_1 + A_1 B_2 + A_2 B_1 - A_2 B_2$. They proved that the expectation value of S must be in the range

$$-2 \leq \langle S \rangle \leq 2 \qquad : \text{hidden-variable theory}, \tag{2}$$

if the local hidden-variable theory is correct. We call (2) the BCHSH inequality abbreviating the names of the authors, Bell, Clauser, Horne, Shimony and Holt. On the other hand, quantum mechanics predicts that

$$-2\sqrt{2} \leq \langle S \rangle \leq 2\sqrt{2} \qquad : \text{quantum mechanics.} \qquad (3)$$

If experiments yield the value of $\langle S \rangle$ in the range $-2\sqrt{2} \leq \langle S \rangle < -2$ or in $2 < \langle S \rangle \leq 2\sqrt{2}$, the BCHSH inequality is violated and we can conclude that the hidden-variable theory is wrong and quantum mechanics is correct. After the proposal of the BCHSH inequality, many experiments have been performed[8-11] and they revealed violations of the BCHSH inequality. Thus, there is no doubt of the failure of the hidden-variable theory.

However, the meaning of the quantity $S = A_1 B_1 + A_1 B_2 + A_2 B_1 - A_2 B_2$ seems obscure. Many generalizations of the BCHSH inequality have been proposed by other researchers[12-14] and a systematic method for generalization has also been given by Avis, Moriyama and Owari,[15] who used the methods of operations research. However, it is still desirable to construct Bell-like inequalities with the understanding of the physical principle that enables the systematic construction.

As another unsatisfactory point of the usual arguments, it should be noted that *the BCHSH inequality is not a fair test for the hidden-variable theory* in a sense explained below. The types of tests of the hidden-variable theory and quantum mechanics are classified as the following. In general, for a physical quantity T, each theory predicts that the expectation value

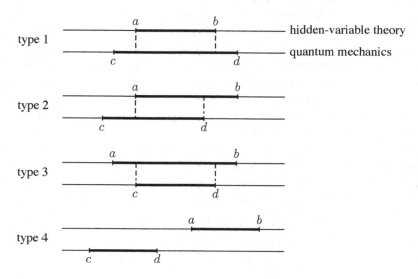

Fig. 1. Classification of tests of the hidden-variable theory and quantum mechanics. Ranges of the predictions of the two theories are indicated by thick bars.

of T falls in some range as

$$a \leq \langle T \rangle \leq b \qquad : \text{hidden-variable theory}, \qquad (4)$$

$$c \leq \langle T \rangle \leq d \qquad : \text{quantum mechanics}. \qquad (5)$$

Then, by measuring the experimental value, we can judge the validities of the two theories. We classify the tests into four types:

type 1: $c < a < b < d$

type 2: $c < a < d < b$, or $a < c < b < d$

type 3: $a < c < d < b$

type 4: $c < d < a < b$, or $a < b < c < d$. $\qquad (6)$

According to this scheme, the BCHSH inequality belongs to the type 1, where the range of prediction of the hidden-variable theory is included in the range of quantum mechanics. Thus, in any experiment of the type 1, it cannot happen that only the hidden-variable theory is correct and quantum mechanics is wrong.

However, if a test of the type 2 with $c < a < d < b$ can be realized, and if the experimental output falls in the range $d < \langle T \rangle \leq b$, we should conclude that the hidden-variable theory is correct and quantum mechanics is wrong.

On the other hand, the Kochen-Specker theorem[16] and the Greenberger-Horne-Shimony-Zeilinger test[17–19] belong to the type 4, where the ranges of the hidden-variable theory and quantum mechanics are completely disjoint.

Of course, we do not expect that any experiments invalidate quantum mechanics in the real world. However, *it is desirable for strengthening the validity of quantum mechanics to provide a test that can reveal even the failure of quantum mechanics and success of the hidden-variable theory.* By passing such a severe test like type 2 or type 4, quantum mechanics will become more persuasive and reliable.

In this paper, we explain several mathematical reasons for the violation of the BCHSH inequality. We also introduce a method for making systematic generalizations of the BCHSH inequality; this method is the main result of this paper. As a product of the main result, we invent a quantity

$$\begin{aligned} T = {} & A_1 B_3 + A_1 B_6 + A_2 B_3 - A_2 B_6 \\ & + A_3 B_2 + A_3 B_5 + A_1 B_2 - A_1 B_5 \\ & + A_2 B_1 + A_2 B_4 + A_3 B_1 - A_3 B_4, \end{aligned} \qquad (7)$$

where the observables A_i ($i = 1, 2, 3$) and B_j ($j = 1, \cdots, 6$) take their values in $\{1, -1\}$ and A_i commutes with B_j. We will show that the two theories predict the range of the expectation value as

$$-6 \le \langle T \rangle \le 6 \qquad : \text{hidden-variable theory,}$$
$$-6\sqrt{2} \le \langle T \rangle \le 2\sqrt{2} \quad : \text{quantum mechanics.} \qquad (8)$$

Hence, this set of inequalities belongs to the type 2, which offers a severer and fairer test for comparing quantum mechanics and the hidden-variable theory than the conventional BCHSH inequality.

2. Bell-Clauser-Horne-Shimony-Holt Inequality

2.1. *Setting up*

The scheme of a Bell-type experiment is briefly sketched in this section for a pedagogical purpose. Of course, this scheme is already well known and hence the reader may skip this part.

Assume that two spin-$\frac{1}{2}$ particles are emitted simultaneously from a common source. The particle pairs are generated repeatedly and they are prepared to be in the same state. The two particles fly in opposite directions, left and right. Two detectors are placed far from the particle source. Each detector receives one particle and measures a component of the spin of the particle. If the spin received by the left detector is in the up state, the value

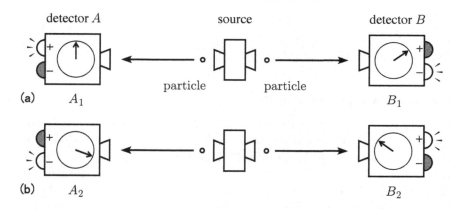

Fig. 2. A setup of Bell-type experiment is depicted à la Mermin.[17,20,21] Two particles are emitted simultaneously from a common source. Each detector has one dial and two lamps. If the spin of the received particle is parallel to the dial, the plus sign lights. If the spin is antiparallel to the dial, the minus sign lights. The adjustable dial angle α of the detector A defines the observable A_α. Similarly, the dial angle β of the detector B defines B_β. (a) $A_1 = +1$, $B_1 = -1$. (b) $A_2 = -1$, $B_2 = +1$.

of an observable A is defined to be $A = +1$. If the spin received by the left detector is in the down state, the value of A is defined to be $A = -1$. Similarly, the spin measured by the right detector defines an observable B, whose value is ± 1 according to up or down state of the detected particle.

It is also assumed that the experimentalist can rearrange the detectors to choose the spin component to be measured. The left detector has an adjustable angle variable α and measures the observable $A(\alpha)$ of the spin component in the direction of α. Similarly, the right one has an angle variable β and measures the spin observable $B(\beta)$. For selected angles $\alpha_1, \alpha_2, \beta_1, \beta_2$, the four observables $A_1 = A(\alpha_1)$, $A_2 = A(\alpha_2)$, $B_1 = B(\beta_1)$ and $B_2 = B(\beta_2)$ are defined. Each observable takes $+1$ or -1 as its value according to up or down state of the spin of the detected particle.

The quantity to be evaluated is the average of

$$S = A_1 B_1 + A_1 B_2 + A_2 B_1 - A_2 B_2. \tag{9}$$

Once the experimentalist fixes the dials at α_1 and β_1, he will measure the outputs A_1, B_1 and calculate the product $A_1 B_1$ for a generated particle pair. This procedure is repeated over particle pairs and the average $\langle A_1 B_1 \rangle$ is determined. In another run, measurements of A_1 and B_2 are repeated, and the average $\langle A_1 B_2 \rangle$ is determined. In a similar way, averages of other terms are also measured and added up for defining

$$\langle S \rangle = \langle A_1 B_1 \rangle + \langle A_1 B_2 \rangle + \langle A_2 B_1 \rangle - \langle A_2 B_2 \rangle. \tag{10}$$

This is the BCHSH quantity.

2.2. *Local hidden-variable theory*

Here we present a quick derivation of the BCHSH inequality (2). In the context of the hidden-variable theory, it is assumed that the state of the pair of particles is uniquely specified by a hidden variable λ. The observable A is assumed to be a function $A(\alpha, \lambda)$ of the parameter α that specifies the configuration of the detector and the hidden variable λ. Once the value of λ is fixed, the measured values of A_1, A_2, B_1, B_2 are fixed to be $+1$ or -1. Actually, it is assumed that the hidden variable is not fixed but it obeys a probability distribution $P(\lambda)$, which is non-negative and normalized. The average of the observable A is defined with the probability $P(\lambda)$ as

$$\langle A_\alpha \rangle = \int A(\alpha, \lambda) P(\lambda) \, d\lambda. \tag{11}$$

Furthermore, if we measure two observables A and B at spatially separated places, the average of product of A and B is calculated as

$$\langle A_\alpha B_\beta \rangle = \int A(\alpha, \lambda) B(\beta, \lambda) P(\lambda) \, d\lambda. \tag{12}$$

This is the scheme of the local hidden-variable theory.

Using the above scheme we can deduce the upper and lower bounds of the quantity

$$S = A_1(B_1 + B_2) + A_2(B_1 - B_2). \tag{13}$$

As stated above, the values of $A_i(\lambda) = A(\alpha_i, \lambda)$ and $B_i(\lambda) = B(\beta_i, \lambda)$ are in $\{+1, -1\}$. When $B_1 + B_2 = \pm 1 \pm 1 = \pm 2$, the other term is $B_1 - B_2 = 0$. When $B_1 + B_2 = \pm 1 \mp 1 = 0$, the other term is $B_1 - B_2 = \pm 2$. Thus, one of $(B_1 + B_2)$ or $(B_1 - B_2)$ is always 0 and the other is ± 2. The coefficients A_1, A_2 are also $+1$ or -1. Hence, the possible values of the quantity $S = A_1(B_1 + B_2) + A_2(B_1 - B_2)$ are ± 2. Since the probability is assumed to be nonnegative and normalized, the average of S is between the minimum and the maximum of S,

$$-2 \leq \langle S \rangle = \int \left(A_1 B_1 + A_1 B_2 + A_2 B_1 - A_2 B_2 \right) P(\lambda) \, d\lambda \leq 2. \tag{14}$$

Thus, the BCHSH inequality (2) is proved.

2.3. Quantum mechanics

Let us turn to quantum mechanics. In the context of quantum mechanics, the state of a pair of spin-$\frac{1}{2}$ particles are described by a vector in the two-qubit Hilbert space $\mathbb{C}^2 \otimes \mathbb{C}^2$ and the observables of the pair are described by operators on $\mathbb{C}^2 \otimes \mathbb{C}^2$. The spin observable measured at the left detector or the right detector is defined to be

$$A(\alpha) = (\sigma_z \cos\alpha + \sigma_x \sin\alpha) \otimes I, \tag{15}$$

$$B(\beta) = I \otimes (\sigma_z \cos\beta + \sigma_x \sin\beta), \tag{16}$$

respectively. Here I is the two-dimensional identity matrix and σ_i are the Pauli matrices. The parameters α and β are adjustable for specifying the spin component to be measured. Substituting functions of a common angle variable θ into them, four observables

$$A_1 = A(0) = \sigma_z \otimes I, \tag{17}$$

$$A_2 = A(2\theta) = (\sigma_z \cos 2\theta + \sigma_x \sin 2\theta) \otimes I, \tag{18}$$

$$B_1 = B(\theta) = I \otimes (\sigma_z \cos\theta + \sigma_x \sin\theta), \tag{19}$$

$$B_2 = B(-\theta) = I \otimes (\sigma_z \cos\theta - \sigma_x \sin\theta) \tag{20}$$

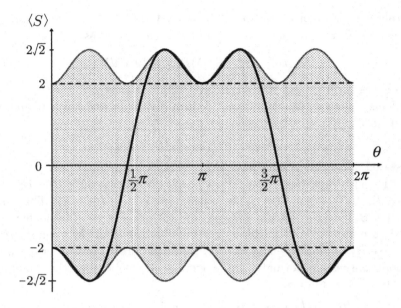

Fig. 3. A test of the ordinary Bell-Clauser-Horne-Shimony-Holt inequality. The hidden-variable theory predicts that $-2 \leq \langle S \rangle \leq 2$. Quantum mechanics predicts that $\langle S \rangle$ is in the gray area B_S, which is defined in (25). The thick curve is the expectation value (28) associated with the singlet state.

are defined. From the matrix representation for S

$$S_\theta = A_1 B_1 + A_1 B_2 + A_2 B_1 - A_2 B_2$$

$$= 2\cos\theta \begin{pmatrix} 1 & 0 & 0 & 0 \\ 0 & -1 & 0 & 0 \\ 0 & 0 & -1 & 0 \\ 0 & 0 & 0 & 1 \end{pmatrix}$$

$$+ 2\cos 2\theta \sin\theta \begin{pmatrix} 0 & 1 & 0 & 0 \\ 1 & 0 & 0 & 0 \\ 0 & 0 & 0 & -1 \\ 0 & 0 & -1 & 0 \end{pmatrix} + 2\sin 2\theta \sin\theta \begin{pmatrix} 0 & 0 & 0 & 1 \\ 0 & 0 & 1 & 0 \\ 0 & 1 & 0 & 0 \\ 1 & 0 & 0 & 0 \end{pmatrix}, \quad (21)$$

the eigenvalues of S_θ are calculated to be

$$\{s_1(\theta), s_2(\theta), s_3(\theta), s_4(\theta)\}$$
$$= \left\{ 2\cos 2\theta, -2\cos 2\theta, 2\sqrt{1 + \sin^2 2\theta}, -2\sqrt{1 + \sin^2 2\theta} \right\}. \quad (22)$$

For a general state including a mixed state, the expectation value of S

is given by a convex combination of the eigenvalues, $\langle S_\theta \rangle = \sum_{i=1}^{4} w_i s_i(\theta)$. Then, the sets of expectation values are put in order as

$$Q_S(\theta) := \left\{ \sum_{i=1}^{4} w_i s_i(\theta) \,\middle|\, w_i \in \mathbb{R}, \, 0 \leq w_i \leq 1, \, \sum_{i=1}^{4} w_i = 1 \right\}$$

$$= \left[-2\sqrt{1 + \sin^2 2\theta}, \, 2\sqrt{1 + \sin^2 2\theta} \right] \subset \mathbb{R}, \tag{23}$$

$$Q_S := \bigcup_{0 \leq \theta \leq 2\pi} Q_S(\theta) = \left[-2\sqrt{2}, \, 2\sqrt{2} \right] \subset \mathbb{R}, \tag{24}$$

$$B_S := \left\{ (\theta, s) \,\middle|\, 0 \leq \theta \leq 2\pi, \, s \in Q_S(\theta) \right\} \subset \mathbb{R}^2. \tag{25}$$

The set $Q_S(\theta)$ is all of possible expectation values of S_θ for a fixed angle θ, and the set Q_S is all the possible expectation values of S_θ for an arbitrary angle θ. The band B_S formed by values allowed by quantum mechanics is shown in Fig. 3 as the shaded domain. It is clear that the set Q_S gives the bound $-2\sqrt{2} \leq \langle S \rangle \leq 2\sqrt{2}$ in the context of quantum mechanics. Thus the inequality (3) is proved.

On the other hand, in the previous subsection, it was proved that the range of the expectation value allowed by the hidden-variable theory is

$$H_S := [-2, 2] \subset \mathbb{R}. \tag{26}$$

Then, it holds that $H_S \subset Q_S(\theta)$ for any θ. Thus, the quantity S provides only the type 1 test.

In particular, if the particle pair is the spin singlet state

$$\psi = \psi_1 + \psi_2 = \frac{1}{\sqrt{2}} \left(|\uparrow\downarrow|-\rangle |\downarrow\uparrow|\rangle \right)$$

$$= \frac{1}{\sqrt{2}} \begin{pmatrix} 1 \\ 0 \end{pmatrix} \otimes \begin{pmatrix} 0 \\ 1 \end{pmatrix} - \frac{1}{\sqrt{2}} \begin{pmatrix} 0 \\ 1 \end{pmatrix} \otimes \begin{pmatrix} 1 \\ 0 \end{pmatrix} = \frac{1}{\sqrt{2}} \begin{pmatrix} 0 \\ 1 \\ -1 \\ 0 \end{pmatrix}, \tag{27}$$

the expectation value becomes

$$\langle \psi | S_\theta | \psi \rangle = -2\cos\theta - 2\sin 2\theta \sin\theta. \tag{28}$$

At $\theta = \pm\frac{\pi}{4}$, it becomes $\langle \psi | S_\theta | \psi \rangle = -2\sqrt{2}$. At $\theta = \pm\frac{3}{4}\pi$, it becomes $\langle \psi | S_\theta | \psi \rangle = 2\sqrt{2}$. Thus, the maximum violation of the BCHSH inequality is attained at these angles.

2.4. *Implication of locality*

Here, we explain the implication of locality. In the context of the hidden-variable theory, locality or nonlocality is formulated as follows. Locality requests that the output of one detector is independent of the arrangement of the other detector that are placed far from the first detector. Namely, in the local theory, the output of the detector A is a function $A(\alpha, \lambda)$. of its arrangement parameter α and the object parameter λ. Similarly, the output of the detector B is a function $B(\beta, \lambda)$. Thus the average of the product quantity is given by the formula

$$\langle A_\alpha B_\beta \rangle = \int A(\alpha, \lambda) B(\beta, \lambda) P(\lambda) \, d\lambda \quad : \text{local theory.} \qquad (29)$$

It is justified to rewrite the quantity $S = A_1 B_1 + A_1 B_2 + A_2 B_1 - A_2 B_2$ defined at (9) into the form $S = A_1(B_1 + B_2) + A_2(B_1 - B_2)$. We can deduce that $B(\beta_1, \lambda) + B(\beta_2, \lambda) = \pm 2$ when $B(\beta_1, \lambda) - B(\beta_2, \lambda) = 0$.

On the other hand, the *nonlocal* hidden-variable theory allows the output of one detector can be affected by the arrangement of the far detector via superluminal signal. Namely, in the nonlocal theory, the output of the detector A can be a function $A(\alpha, \beta, \lambda)$ not only of its arrangement parameter α and the object variable λ, but also of the arrangement parameter β of the distant detector. Similarly, the output of the detector B can be a function $B(\alpha, \beta, \lambda)$. Hence, the above formula (29) is replaced by

$$\langle A_\alpha B_\beta \rangle = \int A(\alpha, \beta, \lambda) B(\alpha, \beta, \lambda) P(\lambda) \, d\lambda \quad : \text{nonlocal theory.} \qquad (30)$$

So the quantity $S = A_1 B_1 + A_1 B_2 + A_2 B_1 - A_2 B_2$ is actually

$$\begin{aligned} S = {} & A(\alpha_1, \beta_1, \lambda) B(\alpha_1, \beta_1, \lambda) + A(\alpha_1, \beta_2, \lambda) B(\alpha_1, \beta_2, \lambda) \\ & + A(\alpha_2, \beta_1, \lambda) B(\alpha_2, \beta_1, \lambda) - A(\alpha_2, \beta_2, \lambda) B(\alpha_2, \beta_2, \lambda). \end{aligned} \qquad (31)$$

Hence, in this context it cannot be justified to rewrite it in the form $S = A_1(B_1 + B_2) + A_2(B_1 - B_2)$ as we did at (13). What is worse, we cannot say that $B(\alpha_1, \beta_1, \lambda) + B(\alpha_1, \beta_2, \lambda) = \pm 2$ even when $B(\alpha_2, \beta_1, \lambda) - B(\alpha_2, \beta_2, \lambda) = 0$. It should be emphasized that the derivation of the BCHSH inequality, (12)-(14), is based on the *local* hidden-variable theory.

In the context of quantum mechanics, we adopt the conventional interpretation, which tells that the locality implies the commutativity of observables separated by a spacelike distance. The operator A act on the Hilbert space of the particle flying to the left, while the operator B act on the Hilbert space of the particle flying to the right. Thus, they are acting on different spaces, and hence, they commute.

It should be mentioned that commutativity is not a necessary condition for locality. Deutsch[22] constructed a model in which spacelikely separated observables do not commute and showed that no physical pathologies arise in his model. Hence, identification of the locality with the commutativity is still controversial.

3. Why Is the BCHSH Inequality Violated?

3.1. *Interference effect*

In this section, we give several explanations for the violation of the BCHSH inequality. Here, we consider in the context of quantum mechanics.

One of the explanations for the violation is given in terms of the interference effect. If the system is in a mixed state described by the density matrix

$$\rho = |\psi_1\rangle \langle \psi_1| + |\psi_2\rangle \langle \psi_2|$$

$$= \frac{1}{2} \begin{pmatrix} 1 & 0 \\ 0 & 0 \end{pmatrix} \otimes \begin{pmatrix} 0 & 0 \\ 0 & 1 \end{pmatrix} + \frac{1}{2} \begin{pmatrix} 0 & 0 \\ 0 & 1 \end{pmatrix} \otimes \begin{pmatrix} 1 & 0 \\ 0 & 0 \end{pmatrix} = \frac{1}{2} \begin{pmatrix} 0 & 0 & 0 & 0 \\ 0 & 1 & 0 & 0 \\ 0 & 0 & 1 & 0 \\ 0 & 0 & 0 & 0 \end{pmatrix}, \qquad (32)$$

the expectation value of S_θ becomes

$$\mathrm{Tr}\,(S_\theta\,\rho) = \langle \psi_1|S_\theta|\psi_1\rangle + \langle \psi_2|S_\theta|\psi_2\rangle = -2\cos\theta, \qquad (33)$$

and it stays in the range $-2 \leq \mathrm{Tr}\,(S_\theta\,\rho) \leq 2$ being consistent with the BCSHS inequality. Instead of the mixed state, if the pure state $\rho = |\psi\rangle\langle\psi|$ with (27) is substituted, the off-diagonal elements of S_θ in (21) contribute to the expectation value as

$$\langle \psi|S_\theta|\psi\rangle = \langle \psi_1|S_\theta|\psi_1\rangle + \langle \psi_2|S_\theta|\psi_2\rangle + \langle \psi_1|S_\theta|\psi_2\rangle + \langle \psi_2|S_\theta|\psi_1\rangle$$

$$= -2\cos\theta - 2\sin 2\theta \sin\theta. \qquad (34)$$

The cross terms $\langle \psi_1|S_\theta|\psi_2\rangle + \langle \psi_2|S_\theta|\psi_1\rangle$ represent the interference effect, which causes the violation of the BCHSH inequality. The interference of the two terms in the superposed state $\psi = \frac{1}{\sqrt{2}}(|\uparrow\downarrow\,|-\rangle|\downarrow\uparrow\,|\rangle)$ is also called entanglement effect. This kind of explanation for the violation of the BCHSH inequality can be found in recent textbooks.[23]

3.2. *Noncommutativity and a useful trick*

Another explanation for the violation of the BCHSH inequality is given in terms of the noncommutativity of quantum observables. A basic fact underlying the violation is that *the eigenvalues of a sum of noncommutative operators are not equal to the sum of the eigenvalues of the respective operators in general.* Let us observe this fact via a simple example. The eigenvalues of the two Pauli matrices

$$\sigma_x = \begin{pmatrix} 0 & 1 \\ 1 & 0 \end{pmatrix}, \qquad \sigma_z = \begin{pmatrix} 1 & 0 \\ 0 & -1 \end{pmatrix} \tag{35}$$

are ± 1. Does the matrix $C := \sigma_x + \sigma_z$ have ± 2 or 0 as its eigenvalues? *No!* Through the calculation

$$\det(sI - C) = \det \begin{pmatrix} s-1 & -1 \\ -1 & s+1 \end{pmatrix} = s^2 - 2, \tag{36}$$

it is easily seen that eigenvalues of C are $\pm\sqrt{2}$. We may say symbolically that *in quantum mechanics 1+1 is not 2 but $\sqrt{2}$.*

This kind of phenomenon can happen since noncommuting observables do not have certain values simultaneously. This property of the noncommuting matrices is summarized as a useful trick that is written in the typical form as

$$\frac{1}{\sqrt{2}}(\sigma_z + \sigma_x) + \frac{1}{\sqrt{2}}(\sigma_z - \sigma_x) = \sqrt{2}\,\sigma_z. \tag{37}$$

Each term $\frac{1}{\sqrt{2}}(\sigma_z \pm \sigma_x)$ in the left-hand side has the spectrum $\{1, -1\}$. Hence, a naive application of the arithmetic rule tells that the value of the left-hand side should be in $\{2, 0, -2\}$. But, actually the right term $\sqrt{2}\,\sigma_z$ has the spectrum $\{\sqrt{2}, -\sqrt{2}\}$. This discrepancy is the basic mechanism that causes the violation of the BCHSH inequality.

The hidden-variable theory assumes that the observables A_α and B_β have certain values $A_\alpha(\lambda)$ and $B_\beta(\lambda)$ at any time even when the observables are not measured. It also assumes that the values obey the ordinary arithmetic rule. The reasoning based on these assumptions leads to the BCHSH inequality (12). However, in quantum mechanics, values cannot be assigned to the noncommuting observables simultaneously and the naive arithmetic rule is not applicable to their values. Hence, the BCHSH inequality is violated.

This type of reasoning reveals the trick for making the BCHSH quantity S. The starting point is

$$S = \sqrt{2}\,(\sigma_z \otimes \sigma_z + \sigma_x \otimes \sigma_x). \tag{38}$$

Note that the spectra of $\sigma_z \otimes \sigma_z$ and $\sigma_x \otimes \sigma_x$ are both $\{1, -1\}$. Since $\sigma_z \otimes \sigma_z$ and $\sigma_x \otimes \sigma_x$ commute, the naive arithmetic rule is applicable to them and the spectrum of S should be a subset of $\{2\sqrt{2}, 0, -2\sqrt{2}\}$. Actually, the spectrum of S is $\{2\sqrt{2}, -2\sqrt{2}\}$. Then, by applying the trick (37) for rewriting S, we obtain

$$
\begin{aligned}
S &= \sqrt{2}\,(\sigma_z \otimes \sigma_z + \sigma_x \otimes \sigma_x) \\
&= \sqrt{2}\Big[\sigma_z \otimes \frac{1}{2}\{(\sigma_z + \sigma_x) + (\sigma_z - \sigma_x)\} + \sigma_x \otimes \frac{1}{2}\{(\sigma_z + \sigma_x) - (\sigma_z - \sigma_x)\}\Big] \\
&= \sigma_z \otimes \frac{1}{\sqrt{2}}(\sigma_z + \sigma_x) + \sigma_z \otimes \frac{1}{\sqrt{2}}(\sigma_z - \sigma_x) \\
&\quad + \sigma_x \otimes \frac{1}{\sqrt{2}}(\sigma_z + \sigma_x) - \sigma_x \otimes \frac{1}{\sqrt{2}}(\sigma_z - \sigma_x) \\
&= A_1 B_1 + A_1 B_2 + A_2 B_1 - A_2 B_2,
\end{aligned}
\tag{39}
$$

which is the quantity maximally violating the BCHSH inequality. By introducing the adjustable parameter θ, we obtain the quantity S_θ expressed in terms of the observables (17)–(20)

Actually, Bell[5,19] himself noticed that the values of noncommuting observables do not obey the naive additive law but he did not utilize this property to derive his inequality. Seevinck and Uffink[24] argued that the noncommutativity is related to the violation of the original BCHSH inequality, but they did not consider a method for generating variations of the BCHSH inequality.

4. New Test

4.1. New quantity

Here, we build a new quantity T that satisfies a new type of the Bell-like inequality. Our starting point is the quantity

$$
T = 2\sqrt{2}\,(\sigma_x \otimes \sigma_x + \sigma_y \otimes \sigma_y + \sigma_z \otimes \sigma_z).
\tag{40}
$$

The three terms $\sigma_a \otimes \sigma_a$ $(a = x, y, z)$ are mutually commutative and their spectra are $\{1, -1\}$. Therefore, the spectrum of T should be a subset of $\{6\sqrt{2}, 2\sqrt{2}, -2\sqrt{2}, -6\sqrt{2}\}$. The matrix representation of T is calculated as

$$\frac{1}{2\sqrt{2}} T = \begin{pmatrix} 0 & 1 \\ 1 & 0 \end{pmatrix} \otimes \begin{pmatrix} 0 & 1 \\ 1 & 0 \end{pmatrix} + \begin{pmatrix} 0 & -i \\ i & 0 \end{pmatrix} \otimes \begin{pmatrix} 0 & -i \\ i & 0 \end{pmatrix} + \begin{pmatrix} 1 & 0 \\ 0 & -1 \end{pmatrix} \otimes \begin{pmatrix} 1 & 0 \\ 0 & -1 \end{pmatrix}$$

$$= \begin{pmatrix} 0 & 0 & 0 & 1 \\ 0 & 0 & 1 & 0 \\ 0 & 1 & 0 & 0 \\ 1 & 0 & 0 & 0 \end{pmatrix} + \begin{pmatrix} 0 & 0 & 0 & -1 \\ 0 & 0 & 1 & 0 \\ 0 & 1 & 0 & 0 \\ -1 & 0 & 0 & 0 \end{pmatrix} + \begin{pmatrix} 1 & 0 & 0 & 0 \\ 0 & -1 & 0 & 0 \\ 0 & 0 & -1 & 0 \\ 0 & 0 & 0 & 1 \end{pmatrix}$$

$$= \begin{pmatrix} 1 & 0 & 0 & 0 \\ 0 & -1 & 2 & 0 \\ 0 & 2 & -1 & 0 \\ 0 & 0 & 0 & 1 \end{pmatrix}. \tag{41}$$

It is easily seen that the eigenvalues of T are $2\sqrt{2}$ (threefold degeneracy) and $-6\sqrt{2}$ (no degeneracy); the corresponding eigenvectors are

$$2\sqrt{2}: a \begin{pmatrix} 1 \\ 0 \\ 0 \\ 0 \end{pmatrix} + b \begin{pmatrix} 0 \\ 1 \\ 1 \\ 0 \end{pmatrix} + c \begin{pmatrix} 0 \\ 0 \\ 0 \\ 1 \end{pmatrix}, \qquad -6\sqrt{2}: \begin{pmatrix} 0 \\ 1 \\ -1 \\ 0 \end{pmatrix}. \tag{42}$$

In other words, the triplet spin states have the eigenvalue $T = 2\sqrt{2}$, while the singlet spin state has the eigenvalue $T = -6\sqrt{2}$. Hence, quantum mechanics predicts the bound

$$-6\sqrt{2} \leq \langle T \rangle \leq 2\sqrt{2} \qquad : \text{quantum mechanics} \tag{43}$$

of the expectation value for any state.

Next, by applying the trick (37) several times, we rewrite T as

$$T = 2\sqrt{2}\,(\sigma_x \otimes \sigma_x + \sigma_y \otimes \sigma_y + \sigma_z \otimes \sigma_z)$$

$$= \frac{2\sqrt{2}}{4} \Big[\sigma_x \otimes \{(\sigma_x + \sigma_y) + (\sigma_x - \sigma_y) + (\sigma_z + \sigma_x) - (\sigma_z - \sigma_x)\}$$

$$+ \sigma_y \otimes \{(\sigma_y + \sigma_z) + (\sigma_y - \sigma_z) + (\sigma_x + \sigma_y) - (\sigma_x - \sigma_y)\}$$

$$+ \sigma_z \otimes \{(\sigma_z + \sigma_x) + (\sigma_z - \sigma_x) + (\sigma_y + \sigma_z) - (\sigma_y - \sigma_z)\} \Big]. \tag{44}$$

By introducing

$$A_1 = \sigma_x \otimes I, \qquad A_2 = \sigma_y \otimes I, \qquad A_3 = \sigma_z \otimes I,$$

$$B_1 = \frac{1}{\sqrt{2}} I \otimes (\sigma_y + \sigma_z), \qquad B_2 = \frac{1}{\sqrt{2}} I \otimes (\sigma_z + \sigma_x),$$

$$B_3 = \frac{1}{\sqrt{2}} I \otimes (\sigma_x + \sigma_y), \qquad B_4 = \frac{1}{\sqrt{2}} I \otimes (\sigma_y - \sigma_z),$$

$$B_5 = \frac{1}{\sqrt{2}} I \otimes (\sigma_z - \sigma_x), \qquad B_6 = \frac{1}{\sqrt{2}} I \otimes (\sigma_x - \sigma_y), \tag{45}$$

we reach the expression

$$\begin{aligned}
T &= A_1(B_3 + B_6 + B_2 - B_5) + A_2(B_1 + B_4 + B_3 - B_6) \\
&\quad + A_3(B_2 + B_5 + B_1 - B_4) \\
&= A_1(B_3 + B_6) + A_2(B_3 - B_6) \\
&\quad + A_3(B_2 + B_5) + A_1(B_2 - B_5) \\
&\quad + A_2(B_1 + B_4) + A_3(B_1 - B_4),
\end{aligned} \tag{46}$$

which was shown at (7) in the Introduction.

4.2. *Hidden-variable theory vs. quantum mechanics*

The hidden-variable theory is applicable to T in this form (46). In the context of the hidden-variable theory, A_α and B_β are functions of the hidden variable λ, and the values of $A_\alpha(\lambda)$ and $B_\beta(\lambda)$ are in $\{1, -1\}$. Then, it is easily seen from the expression (46) that the possible values of $T(\lambda)$ are in $\{6, 2, -2, -6\}$. Hence, the average $\langle T \rangle = \int T(\lambda)P(\lambda)d\lambda$ is in the range

$$-6 \leq \langle T \rangle \leq 6 \qquad : \text{hidden-variable theory.} \tag{47}$$

The range (43) of the prediction of quantum mechanics and the range (47) of the hidden-variable theory have the overlap $[-6, 2\sqrt{2}]$ where both the two theories hold, and the region $[-6\sqrt{2}, -6) \cup (2\sqrt{2}, 6]$ where only one of the two theories holds. Thus, T is a quantity that realizes the type 2 test.

Moreover, we introduce a parameter θ and define T_θ as the polynomial

(46) of

$$A_1 = (\sigma_z \cos 2\theta + \sigma_x \sin 2\theta) \otimes I, \tag{48}$$

$$A_2 = (\sigma_z \cos 2\theta + \sigma_y \sin 2\theta) \otimes I, \tag{49}$$

$$A_3 = \sigma_z \otimes I, \tag{50}$$

$$B_1 = I \otimes (\sigma_y \cos \theta + \sigma_z \sin \theta), \tag{51}$$

$$B_2 = I \otimes (\sigma_z \cos \theta + \sigma_x \sin \theta), \tag{52}$$

$$B_3 = I \otimes (\sigma_x \cos \theta + \sigma_y \sin \theta), \tag{53}$$

$$B_4 = I \otimes (\sigma_y \cos \theta - \sigma_z \sin \theta), \tag{54}$$

$$B_5 = I \otimes (\sigma_z \cos \theta - \sigma_x \sin \theta), \tag{55}$$

$$B_6 = I \otimes (\sigma_x \cos \theta - \sigma_y \sin \theta). \tag{56}$$

The parameter θ is interpreted as an angle that specifies the directions of the detectors. The matrix representation of T_θ is

$$T_\theta = 2(\cos \theta + \sin \theta) \times$$
$$\begin{pmatrix} 1 & (1-i)\cos 2\theta & 0 & 0 \\ (1+i)\cos 2\theta & -1 & 2\sin 2\theta & 0 \\ 0 & 2\sin 2\theta & -1 & -(1-i)\cos 2\theta \\ 0 & 0 & -(1+i)\cos 2\theta & 1 \end{pmatrix}. \tag{57}$$

When $\theta = \frac{\pi}{4}$, T_θ is reduced to the original form (41). The eigenvalues of T_θ are

$$\{t_1(\theta), t_2(\theta), t_3(\theta), t_4(\theta)\}$$
$$= \Big\{ 2(\cos \theta + \sin \theta)\big(-\sin 2\theta \pm \sqrt{\cos^2 2\theta + 2 + 2\sin 2\theta}\,\big),$$
$$2(\cos \theta + \sin \theta)\big(\sin 2\theta \pm \sqrt{\cos^2 2\theta + 2 - 2\sin 2\theta}\,\big) \Big\}. \tag{58}$$

The sets of values allowed by quantum mechanics are arranged as

$$Q_T(\theta) := \Big\{ \sum_{i=1}^{4} w_i t_i(\theta) \,\Big|\, w_i \in \mathbb{R},\, 0 \le w_i \le 1,\, \sum_{i=1}^{4} w_i = 1 \Big\} \subset \mathbb{R}, \tag{59}$$

$$Q_T := \bigcup_{0 \le \theta \le 2\pi} Q_T(\theta) = [-6\sqrt{2},\, 6\sqrt{2}] \subset \mathbb{R}, \tag{60}$$

$$B_T := \{(\theta, t) \,|\, 0 \le \theta \le 2\pi,\, t \in Q_T(\theta)\} \subset \mathbb{R}^2. \tag{61}$$

And the range allowed by the hidden-variable theory is denoted as

$$H_T := [-6,\, 6] \subset \mathbb{R}. \tag{62}$$

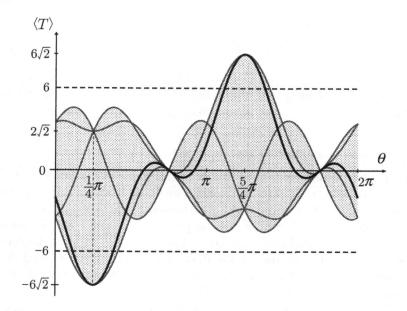

Fig. 4. A test of the new inequality. The hidden-variable theory predicts that $-6 \leq \langle T \rangle \leq 6$. Quantum mechanics predicts that $\langle T \rangle$ is in the gray area B_T, which is defined in (61). The thick curve is the expectation value (63) associated with the singlet state.

In Fig. 4, the band B_T is shown as the shaded region. At any fixed value of θ, the predictions of the two theories have some overlap, namely, we have $Q_T(\theta) \cap H_T \neq \varnothing$. It also happens that $Q_T(\theta) \subset H_T$ for some θ. However, it never happens that $Q_T(\theta) \supset H_T$ at any θ. In this sense, the tests of type 2 and type 3 are realized. It is also to be noted that $Q_T \supset H_T$, namely, the whole set of quantum predictions is wider than the predictions of the hidden-variable theory.

In particular, the singlet state ψ of (27) yields the expectation value

$$f(\theta) = \langle \psi | T_\theta | \psi \rangle = -2(\cos \theta + \sin \theta)(1 + 2 \sin 2\theta). \qquad (63)$$

The range $F := \{ f(\theta) \,|\, 0 \leq \theta \leq 2\pi \} = [-6\sqrt{2}, 6\sqrt{2}]$ covers H_T completely. This means that the violation of the bound of the hidden-variable theory are observed with the state ψ.

If we take another entangled state

$$\chi = \frac{1}{\sqrt{2}} \big(|\mathord{\uparrow\uparrow} \,|+\rangle\, |\mathord{\downarrow\downarrow}\,|\rangle \big) = \frac{1}{\sqrt{2}} \begin{pmatrix} 1 \\ 0 \end{pmatrix} \otimes \begin{pmatrix} 1 \\ 0 \end{pmatrix} + \frac{1}{\sqrt{2}} \begin{pmatrix} 0 \\ 1 \end{pmatrix} \otimes \begin{pmatrix} 0 \\ 1 \end{pmatrix} = \frac{1}{\sqrt{2}} \begin{pmatrix} 1 \\ 0 \\ 0 \\ 1 \end{pmatrix} \qquad (64)$$

instead of the singlet state ψ, it yields

$$g(\theta) = \langle \chi | T_\theta | \chi \rangle = 2(\cos\theta + \sin\theta). \qquad (65)$$

The range $G := \{g(\theta) \,|\, 0 \le \theta \le 2\pi\} = [-2\sqrt{2},\, 2\sqrt{2}]$ is included in H_T completely. This means that the violation of the hidden-variable theory will never be observed in the measurement of T with the state χ.

5. Summary and Discussion

Here, we give a summary of this study. In this paper, we pointed out that the violation of the BCHSH inequality can be understood as a result of the noncommutativity of quantum observables. For noncommuting operators, an eigenvalue of a sum of operators does not coincide with a sum of eigenvalues of the respective operators. Using this property, we invented a method of building systematically Bell-like observables and showed that the conventional BCHSH inequality is reconstructed by this method. This diminished the ad hoc nature of the BCHSH observable.

We classified possible tests of the hidden-variable theory and quantum mechanics. We pointed out that there was no chance in the conventional BCHSH test to reveal the invalidity of quantum mechanics with the validity of the hidden-variable theory. By applying our method, we constructed the new observable T and calculated the range of its average in the contexts of the hidden-variable theory and quantum mechanics, respectively. It was shown that there is a chance that the new test with T reveals the invalidity of quantum mechanics with the validity of the hidden-variable theory.

Of course, we do not aim to deny the validity of quantum mechanics, but we aim to support it by passing the new severer test. It is also our purpose to clarify the implication of violations of Bell-like inequalities of various types.

There are several remaining problems concerning the generalized inequality. The first one is the existence problem in a mathematical sense. The Bell-like inequality is a necessary condition for the existence of the probability distribution that satisfies (29). In the case of the conventional BCHSH inequality, Fine[25] proved that the set of inequalities

$$-2 \le \langle A_1 B_1 \rangle + \langle A_1 B_2 \rangle + \langle A_2 B_1 \rangle - \langle A_2 B_2 \rangle \le 2, \qquad (66)$$

$$-2 \le \langle A_1 B_2 \rangle + \langle A_2 B_1 \rangle + \langle A_2 B_2 \rangle - \langle A_1 B_1 \rangle \le 2, \qquad (67)$$

$$-2 \le \langle A_2 B_1 \rangle + \langle A_2 B_2 \rangle + \langle A_1 B_1 \rangle - \langle A_1 B_2 \rangle \le 2, \qquad (68)$$

$$-2 \le \langle A_2 B_2 \rangle + \langle A_1 B_1 \rangle + \langle A_1 B_2 \rangle - \langle A_2 B_1 \rangle \le 2 \qquad (69)$$

is a necessary and sufficient condition for the existence of the probability distribution of the hidden variable. Although there are studies on the tightness of some variations of the BCHSH inequalities,[26–28] finding the necessary and sufficient condition for the existence of the probability distribution of the hidden variable for our observable T is left as an open problem.

The second one is a practical problem. In principle, the test that we proposed can be implemented in an experiment using pairs of photons or spin-half particles. But our choice involves nine observables (48)–(56) with one variable angle. Hence, our scheme is still too cumbersome for practical use. It is more desirable to reduce the number of observables to make the experiments easier.

The third one is the extensibility of our scheme. The proposed observable T is defined in the two-qubit Hilbert space. It is possible to extend our scheme to multiqubit systems. It is also desirable to construct a Bell-like quantity with less number of observables.

Acknowledgments

I thank Prof. M. Nakahara for giving me the opportunity of this presentation. This work was supported by a Grant-in-Aid for Scientific Research of the Japan Society for the Promotion of Science (Grant No. 22540410).

References

1. T. Isobe and S. Tanimura, *Prog. Theor. Phys.* **124**, 191 (2010).
2. D. Bohm, *Phys. Rev.* **85**, 166 (1952).
3. For a review, see F. J. Belinfante, *A Survey of Hidden-Variables Theories* (Pergamon Press, 1973).
4. J. S. Bell, *Physics* **1**, 195 (1964).
5. J. S. Bell, *Rev. Mod. Phys.* **38**, 447 (1966).
6. J. F. Clauser, M. A. Horne, A. Shimony and R. A. Holt, *Phys. Rev. Lett.* **23**, 880 (1969).
7. J. F. Clauser and M. A. Horne, *Phys. Rev. D* **10**, 526 (1974).
8. J. F. Clauser and A. Shimony, *Rep. Prog. Phys.* **41**, 1881 (1978).
9. A. Aspect, P. Grangier and G. Roger, *Phys. Rev. Lett.* **47**, 460 (1981).
10. A. Aspect, P. Grangier and G. Roger, *Phys. Rev. Lett.* **49**, 91 (1982).
11. H. Sakai et al., *Phys. Rev. Lett.* **97**, 150405 (2006).
12. M. Froissart, *Il Nuovo Cim. B* **64**, 241 (1981).
13. A. Peres, *Found. Phys.* **29**, 589 (1999).
14. D. Collins and N. Gisin, *J. Phys. A* **37**, 1775 (2004).
15. D. Avis, S. Moriyama and M. Owari, *IEICE Trans. Fundamentals E* **92-A**, 1254 (2009).

16. S. Kochen and E. P. Specker, *J. Math. Mech.* **17**, 59 (1967).
17. D. Mermin, *Am. J. Phys.* **58**, 731 (1990).
18. D. M. Greenberger, M. A. Horne, A. Shimony and A. Zeilinger, *Am. J. Phys.* **58**, 1131 (1990).
19. D. Mermin, *Rev. Mod. Phys.* **65**, 803 (1993).
20. N. D. Mermin *Am. J. Phys.* **49**, 940 (1981).
21. N. D. Mermin, *Boojums All the Way Through* (Cambridge University Press, 1990).
22. D. Deutsch, e-print arXiv: quant-ph/0401024v1.
23. For example, see p. 229 of A. Shimizu, *Foundations of Quantum Theory* (in Japanese, *Ryoshiron-no Kiso*), section 8.7 in the revised edition (Saiensu-sha, 2004).
24. M. Seevinck and J. Uffink, *Phys. Rev. A* **76**, 042105 (2007).
25. A. Fine, *Phys. Rev. Lett.* **48**, 291 (1982). [Comments: A. Garg and N. D. Mermin, *Phys. Rev. Lett.* **49**, 242 (1982). A. Fine, *Phys. Rev. Lett.* **49**, 243 (1982).]
26. A. Garg and N. D. Mermin, *Found. Physics* **14**, 1 (1984).
27. W. Laskowski, T. Paterek, M. Żukowski and Č. Brukner, *Phys. Rev. Lett.* **93**, 200401 (2004).
28. K. F. Pál and T. Vértesi, *Phys. Rev. A* **79**, 022120 (2009).

HYBRID QUANTUM ANNEALING FOR CLUSTERING PROBLEMS

SHU TANAKA

Research Center for Quantum Computing, Interdisciplinary Graduate School of Science and Engineering, Kinki University, 3-4-1 Kowakae, Higashi-Osaka, Osaka 577-8502, Japan
E-mail: shu-t@chem.s.u-tokyo.ac.jp

RYO TAMURA

Institute for Solid State Physics, University of Tokyo, 5-1-5 Kashiwanoha, Kashiwa, Chiba 277-0805, Japan
E-mail: r.tamura@issp.u-tokyo.ac.jp

ISSEI SATO

Information Science and Technology, University of Tokyo 7-3-1, Hongo, Bunkyo-ku, Tokyo 113-0033, Japan
E-mail: sato@r.dl.itc.u-tokyo.ac.jp

KENICHI KURIHARA

Google, 6-10-1, Roppongi, Minato-ku, Tokyo 106-0032, Japan

We develop a hybrid type of quantum annealing in which we control temperature and quantum field simultaneously. We study the efficiency of proposed quantum annealing and find a good schedule of changing thermal fluctuation and quantum fluctuation. In this paper, we focus on clustering problems which are important topics in information science and engineering. We obtain the better solution of the clustering problem than the standard simulated annealing by proposed quantum annealing.

Keywords: Quantum annealing; Potts model; Path-integral representation; Monte Carlo simulation; Statistical physics

1. Introduction

Optimization problems have spread in wide area of science, for example, information science and statistical physics. Properties of optimization problems can be summarized as follows:

- There are many elements.

- A cost function can be defined. The best solution is the state where the cost function takes the maximum value or the minimum value depending on the definition of the given problem.

An well-known example of optimization problem is the so-called "traveling salesman problem". The traveling salesman problem is to find the shortest path covering all the cities. Here, we can path through each city only once. In this problem, the cost function is the length of path. In general, it is difficult to obtain the best solution by naive approach since the number of candidates of a solution is very huge. Then, in the optimization problem, it is a central issue how to find the best (or better) solution. There are many types of optimization problems. Depending on an optimization problem, there are specific algorithms to solve the problem.

To treat optimization problem in a general way, on the other hand, a couple of approaches have been developed from a viewpoint of statistical physics. The cost function can be regarded as the internal energy in statistical physics. The cost function of the best solution corresponds to the internal energy of the ground state. Kirkpatrick *et al.* proposed a pioneering generic algorithm – simulated annealing.[1,2] Simulated annealing is a method to obtain not so bad solution of optimization problem by decreasing temperature. In high temperature, since the probability distribution at equilibrium state is almost flat, the state can be changed easily. As we decrease temperature gradually, generated probability distribution expects to approach the equilibrium probability distribution at each temperature. Then we can find not so bad solution of problem by simulated annealing. Because of the Geman-Geman's argument,[3] we can succeed to obtain the best solution with the probability unity if we decrease the temperature slow enough. Although there could be better algorithms specific to each optimization problem, simulated annealing is regarded as a stable and generic method because of easy implementation and independency from problems. The purpose of our study is to establish a good general method for optimization problems from a viewpoint of quantum statistical physics.

The organization of this paper is as follows. We first review on quantum annealing which has been believed a good general method to obtain not so bad solution of optimization problems. In section 3, we will review how to implement Monte Carlo simulation. In section 4, we will introduce a model which can treat clustering problems. We also consider quantum effect on this model in this section. In section 5, some remarks on this problem will be shown. In section 6, we will show results obtained by our proposed method. In section 7, we will summarize our study. We will review on the concepts of "invisible fluctuation" in the appendix A.

2. Quantum Annealing

Quantum annealing is expected that it succeeds to obtain the better solution of optimization problems than simulated annealing.[4-11] This method is based on quantum statistical physics. There have been a couple of realization methods of quantum annealing: (i) stochastic method, (ii) deterministic method, and (iii) experiment on artificial lattice.

The stochastic method is realized by quantum Monte Carlo method in many circumstances.[6,12] The quantum Monte Carlo method is an established method to obtain equilibrium properties of strongly correlated quantum systems. Efficient algorithms for quantum annealing have been developed such as the cluster algorithm.[13,14] Owing to these masterly methods, quantum Monte Carlo simulation can be adopted for large-scale systems.

There are a couple of deterministic methods for quantum annealing. The first one is based on the time-dependent Schrödinger equation.[4] This method can trace the real-time evolution which can be observed in real experiments. This method is called as "quantum adiabatic evolution" in quantum information science.[5] This method cannot treat large-scale systems because of the limitation of size of memory in computer. The second one is based on a time-dependent density matrix renormalization group.[15,16] By this method, we can study time-evolution of one-dimensional quantum systems. However it is difficult to treat two-dimensional or three-dimensional quantum systems by this method. Both of two methods are based on the principle of quantum mechanics. However, it is not necessary to treat the Schrödinger equation directly if we adopt the deterministic method as quantum annealing. This is because our purpose is to obtain not so bad solution of given problem. One of the examples is mean-field calculation.[17,18] This method can treat large-scale systems as well as the quantum Monte Carlo simulation. Then, this method has been widely adopted for optimization problems.

Next there are a couple of proposals for experiments on artificial lattice for quantum annealing. For example, we can generate quantum state by optical lattice.[19-22] In addition, the Ising model with transverse field can be realized by superconducting flux qubits.[23-26] These methods are expected as a new type of quantum computer.

There have been a number of studies on quantum annealing from a viewpoint of theoretical physics. Convergence theorem for quantum annealing was proved as well as that for simulated annealing.[27-29] According to this theorem, the permitted upper bound of sweeping speed of quantum field in quantum annealing is larger than that of temperature in simulated annealing. From this theorem, quantum annealing seems better than simulated annealing in principle. However, we often sweep temperature and/or quantum field faster than that upper bound in practice. Then, it is nontrivial whether quantum annealing is better than simulated annealing from a viewpoint of practical situation. Microscopic behavior in quantum annealing has been also investigated. It is well-known that frustrated systems have some interesting static behavior induced by thermal fluctuation and quantum fluctuation. Dynamical properties of frustrated systems have been studied from a viewpoint of quantum annealing.[30–35] Novel type of implementation of quantum annealing itself is an important topic. One of the examples is quantum annealing based on the Jarzynski equality.[36] This method is expected an efficient method since it uses both merits which come from thermal fluctuation and quantum fluctuation.

In this paper, we adopt the first strategy – the quantum Monte Carlo method as the realization of quantum annealing.

3. Monte Carlo Simulation

In this section, we review how to implement Monte Carlo simulation. It has been often used in order to obtain the equilibrium properties of strongly correlated systems such as magnetic systems and bosonic systems. Equilibrium physical quantities of the system which is expressed by the Hamiltonian $\mathcal{H}$ at finite temperature T is given as

$$\langle \mathcal{O} \rangle_{\text{eq}}^{(T)} = \frac{\text{Tr}\, \mathcal{O} e^{-\beta \mathcal{H}}}{\text{Tr}\, e^{-\beta \mathcal{H}}}, \tag{1}$$

where β denotes the inverse temperature $1/T$ and here the Boltzmann constant k_{B} is set to be unity. If we consider a small system, we can obtain all of the equilibrium physical quantities by naive method. If we consider large scale systems, however, we cannot obtain the equilibrium properties by naive method in practice. Monte Carlo simulation enables us to calculate equilibrium physical quantities with high accuracy by the following relation.

$$\frac{\sum_{\Sigma} \mathcal{O}(\Sigma) e^{-\beta \mathcal{H}(\Sigma)}}{\sum_{\Sigma} e^{-\beta \mathcal{H}(\Sigma)}} \rightarrow \langle \mathcal{O} \rangle_{\text{eq}}^{(T)}, \tag{2}$$

where Σ denotes sample, in other words, state. Physical quantity converges to the equilibrium value as the number of samples increases. In fact, the above calculation is inefficient if states are generated by uniform distribution. In order to make the method more efficient, we generate states according to equilibrium probability distribution which is proportional to the

Boltzmann factor $e^{-\beta E(\Sigma)}$. This method is called the importance sampling. We can obtain the equilibrium value as follows:

$$\frac{\sum_\Sigma \mathcal{O}(\Sigma)}{\sum_\Sigma} \to \langle \mathcal{O} \rangle_{\text{eq}}^{(T)}. \tag{3}$$

In order to generate a state from the equilibrium distribution, we just have to use Markov chain Monte Carlo method. Time evolution of probability distribution is given as the master equation:

$$P(\Sigma_i, t + \Delta t) = -\sum_{j \neq i} P(\Sigma_i, t) w(\Sigma_j | \Sigma_i) \Delta t$$
$$+ \sum_{j \neq i} P(\Sigma_j, t) w(\Sigma_i | \Sigma_j) \Delta t + P(\Sigma_i, t) w(\Sigma_i | \Sigma_i) \Delta t, \tag{4}$$

where $P(\Sigma_i, t)$ denotes probability of the state Σ_i at time t and $w(\Sigma_j | \Sigma_i)$ represents transition probability from the state Σ_i to the state Σ_j in unit time. Transition probability $w(\Sigma_j | \Sigma_i)$ obeys

$$\sum_{\Sigma_j} w(\Sigma_j | \Sigma_i) = 1 \quad (\forall \Sigma_i). \tag{5}$$

The master equation can be represented as

$$\mathbf{P}(t + \Delta t) = \mathcal{L} \mathbf{P}(t), \tag{6}$$

where $\mathbf{P}(t)$ is a vector-representation of probability distribution $\{P(\Sigma_i, t)\}$ and $\mathcal{L}$ is called the transition matrix whose elements are expressed as

$$\mathcal{L}_{ji} = w(\Sigma_j | \Sigma_i) \Delta t, \quad \mathcal{L}_{ii} = 1 - \sum_{j \neq i} \mathcal{L}_{ji} = 1 - \sum_{j \neq i} w(\Sigma_j | \Sigma_i) \Delta t. \tag{7}$$

It should be noted that $\mathcal{L}$ is a non-negative matrix by the definition. This time evolution is the Markovian since the time-evolution operator $\mathcal{L}$ does not depend on time. If the time-evolution operator $\mathcal{L}$ obeys (i) detailed balance condition and (ii) ergodicity, we can obtain the equilibrium probability distribution in the limit of $t \to \infty$ because of the Perron-Frobenius theorem.

4. Model

Clustering problem is one of the important problems in information science and engineering. Since it is difficult to obtain the best solution of the clustering problem by naive method, development of a new method which can obtain the best (or not so bad) solution is an important issue. We can obtain not so bad solution by using simulated annealing as we mentioned in the section 1. We propose a new type of quantum annealing and succeed to obtain better solution by proposed quantum annealing method as will be mentioned.

In the beginning of this section, we will explain a model to consider clustering problems. After that, we will introduce new kind of fluctuation – quantum fluctuation – into this model. Next, we will review on implementation method of the quantum annealing method.

4.1. *Clustering problem*

In clustering problems, there are N elements in the space, which is depicted in Fig. 1(a).

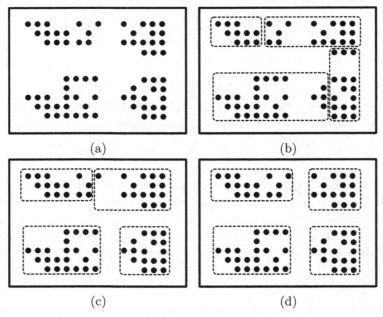

Fig. 1. The dots represent elements. The dotted boxes denote clusters. In these figures, the number of elements $N = 69$ and the number of clusters $Q = 4$. (a) There are N elements in the space. (b) Not so good solution. (c) Not so bad solution. (d) The best solution.

Clustering problems are to decide which the best partition of these elements. In other words, clustering problems are to search the best division of N elements into Q sub-categories. Clustering problems have been applied for not only natural science but also social science. For instance, divisions of the articles in newspaper in terms of contents and analysis of questionnaire can be considered as clustering problems. Figure 1 (b), (c), and (d) represent not so good, not so bad, and the best solution, respectively. In practice, it is difficult to obtain the best solution by direct method for huge number N, since the number of total states is Q^N. Then we often use simulated annealing to solve clustering problems as one of useful generic methods. However, even if we use simulated annealing, it is difficult to obtain the best solution. The energy landscape of clustering problems is very complicated such as random spin systems and frustrated systems. Then we should develop more efficient algorithm than the simulated annealing method. Actually, there is a pioneering method to obtain more better solution than simulated annealing method. This method is called exchange method.[37] In the exchange method, we prepare some independent layers where the temperatures are different. We sometimes exchange the states between two layers according to the Boltzmann weight. In this paper, we adopt another strategy, quantum annealing method as more efficient algorithm. We introduce quantum term into this model to implement the quantum annealing in the next section.

4.2. *Quantum fluctuation*

Before introducing a quantum fluctuation, we review on a classical Hamiltonian of the original clustering problem. The classical Hamiltonian is given by

$$\mathcal{H}_c = \text{diag}\left(E(\Sigma_1), E(\Sigma_2), \cdots, E(\Sigma_{Q^N})\right),\tag{8}$$

where $E(\Sigma_i)$ denotes the eigenenergy of i-th state Σ_i. Suppose we consider the case for $Q = 3$ and $N = 2$ as an example. The classical Hamiltonian is given as

$$
\mathcal{H}_c = \begin{pmatrix}
E(\Sigma_1) & 0 & 0 & 0 & 0 & 0 & 0 & 0 & 0 \\
0 & E(\Sigma_2) & 0 & 0 & 0 & 0 & 0 & 0 & 0 \\
0 & 0 & E(\Sigma_3) & 0 & 0 & 0 & 0 & 0 & 0 \\
0 & 0 & 0 & E(\Sigma_4) & 0 & 0 & 0 & 0 & 0 \\
0 & 0 & 0 & 0 & E(\Sigma_5) & 0 & 0 & 0 & 0 \\
0 & 0 & 0 & 0 & 0 & E(\Sigma_6) & 0 & 0 & 0 \\
0 & 0 & 0 & 0 & 0 & 0 & E(\Sigma_7) & 0 & 0 \\
0 & 0 & 0 & 0 & 0 & 0 & 0 & E(\Sigma_8) & 0 \\
0 & 0 & 0 & 0 & 0 & 0 & 0 & 0 & E(\Sigma_9)
\end{pmatrix}, \quad (9)
$$

where the definitions of the states from Σ_1 to Σ_9 are summarized in Table 1. σ_1 and σ_2 in Table 1 represent the states of the first element and the second element, respectively.

Table 1. Definitions of the states from Σ_1 to Σ_9.

	Σ_1	Σ_2	Σ_3	Σ_4	Σ_5	Σ_6	Σ_7	Σ_8	Σ_9
σ_1	1	2	3	1	2	3	1	2	3
σ_2	1	1	1	2	2	2	3	3	3

Next we introduce a quantum fluctuation into this model. Since when $Q = 2$, the classical Hamiltonian is equivalent to that of the Ising model. It is natural to extend transverse field in the Ising spin system as a quantum fluctuation. Then we adopt the following definition as a quantum fluctuation.

$$
\mathcal{H}_q = -\Gamma \sum_{i=1}^{N} \sigma_i^x = -\Gamma \sum_{i=1}^{N} (\hat{1}_Q - \mathbb{E}_Q), \quad (10)
$$

where $\hat{1}_Q$ denotes the matrix whose all elements are unity and $\mathbb{E}_Q$ represents identity matrix. Both $\hat{1}_Q$ and $\mathbb{E}_Q$ are $Q \times Q$ matrices. It should be noted that σ_i^x denotes the x-component of $s = 1/2$ Pauli matrix at the site i when $Q = 2$. Suppose we consider the case for $Q = 3$ and $N = 2$ as the previous example, the quantum part of the Hamiltonian $\mathcal{H}_q$ is given as

$$\mathcal{H}_q = \begin{pmatrix} 0 & -\Gamma & -\Gamma & -\Gamma & 0 & 0 & -\Gamma & 0 & 0 \\ -\Gamma & 0 & -\Gamma & 0 & -\Gamma & 0 & 0 & -\Gamma & 0 \\ -\Gamma & -\Gamma & 0 & 0 & 0 & -\Gamma & 0 & 0 & -\Gamma \\ -\Gamma & 0 & 0 & 0 & -\Gamma & -\Gamma & -\Gamma & 0 & 0 \\ 0 & -\Gamma & 0 & -\Gamma & 0 & -\Gamma & 0 & -\Gamma & 0 \\ 0 & 0 & -\Gamma & -\Gamma & -\Gamma & 0 & 0 & 0 & -\Gamma \\ -\Gamma & 0 & 0 & -\Gamma & 0 & 0 & 0 & -\Gamma & -\Gamma \\ 0 & -\Gamma & 0 & 0 & -\Gamma & 0 & -\Gamma & 0 & -\Gamma \\ 0 & 0 & -\Gamma & 0 & 0 & -\Gamma & -\Gamma & -\Gamma & 0 \end{pmatrix}. \tag{11}$$

When we use quantum Monte Carlo method, all we have to do is to calculate the probability of the state Σ. In the next section, we will show how to calculate the probability of the state Σ by path-integral representation.[38,39]

4.3. Path integral representation

We consider the following Hamiltonian

$$\mathcal{H} = \mathcal{H}_c + \mathcal{H}_q. \tag{12}$$

When the Hamiltonian $\mathcal{H}$ is a diagonal matrix which corresponds to simulated annealing, i.e. $\Gamma = 0$, the probability of the state Σ at finite temperature T is given as

$$p_{SA}(\Sigma; \beta) = \frac{e^{-\beta E(\Sigma)}}{\text{Tr}\,e^{-\beta \mathcal{H}_c}} = \frac{1}{Z}\langle \Sigma | e^{-\beta \mathcal{H}_c} | \Sigma \rangle, \tag{13}$$

where the denominator is called the partition function in statistical physics. The partition function Z is calculated as

$$Z = \text{Tr}\,e^{-\beta \mathcal{H}_c} = \sum_{\Sigma} \langle \Sigma | e^{-\beta \mathcal{H}_c} | \Sigma \rangle = \sum_{\Sigma} e^{-\beta E(\Sigma)}. \tag{14}$$

We can change the state by using the "single-spin-flip" type of heat bath method,

$$p_{SA}^{\text{update}}(\sigma_i = s | \Sigma \setminus \sigma_i) = \frac{e^{-\beta E(\sigma_i = s, \Sigma \setminus \sigma_i)}}{\sum_{s'=1}^{Q} e^{-\beta E(s', \Sigma \setminus \sigma_i)}}, \tag{15}$$

where $\Sigma \setminus \sigma_i$ means $\{\sigma_j | j \neq i\}$ and $p(A|B)$ denotes a conditional probability of A given B. The denominator of Eq. (15) can be calculated where the computational cost is $\mathcal{O}(Q)$.

In similar with the classical case, the probability of the state Σ at finite temperature T and finite quantum field Γ is given as

$$p_{\mathrm{QA}}(\Sigma; \beta, \Gamma) = \frac{\langle \Sigma | e^{-\beta \mathcal{H}} | \Sigma \rangle}{\sum_{\Sigma'} \langle \Sigma' | e^{-\beta \mathcal{H}} | \Sigma' \rangle} = \frac{\langle \Sigma | e^{-\beta \mathcal{H}} | \Sigma \rangle}{Z}. \tag{16}$$

Note that it is difficult to calculate $\langle \Sigma | e^{-\beta \mathcal{H}} | \Sigma \rangle$, since the Hamiltonian including quantum field has off-diagonal elements. For small systems, we can exactly calculate all of the elements $e^{-\beta \mathcal{H}}$ by using the unitary transform. However, we can not obtain them for large system in practice. In order to calculate the probability given by Eq. (16), we should rewrite the numerator of it by path-integral representation. Then we obtain the probability

$$p_{\mathrm{QA}}(\Sigma; \beta, \Gamma) = \frac{1}{Z} \langle \Sigma | \left(e^{-\frac{\beta}{m} \mathcal{H}_c} e^{-\frac{\beta}{m} \mathcal{H}_q} \right)^m | \Sigma \rangle + \mathcal{O}(\frac{1}{m})$$

$$= \frac{1}{Z} \sum_{\Sigma^{(1)'}} \sum_{\Sigma^{(2)}} \cdots \sum_{\Sigma^{(m)'}} \prod_{j=1}^{m} \langle \Sigma^{(j)} | e^{-\frac{\beta}{m} \mathcal{H}_c} | \Sigma^{(j)'} \rangle \langle \Sigma^{(j)'} | e^{-\frac{\beta}{m} \mathcal{H}_q} | \Sigma^{(j+1)} \rangle, \tag{17}$$

where m is called as the Trotter number and $\Sigma^{(1)} = \Sigma^{(m+1)} = \Sigma$ which corresponds to periodic boundary condition along the Trotter axis.

Here we define

$$s(\Sigma^{(j)}, \Sigma^{(j+1)}) := \frac{1}{N} \sum_{i=1}^{N} \delta \left(\sigma_i^{(j)}, \sigma_i^{(j+1)} \right), \tag{18}$$

$$f(\beta, \Gamma) := N \log \left(1 + \frac{Q}{e^{\frac{Q\beta\Gamma}{m}} - 1} \right), \tag{19}$$

where $\sigma_i^{(j)}$ denotes the i-th element on the j-th Trotter layer. Then we can obtain

$$\langle \Sigma^{(j)} | e^{-\frac{\beta}{m} \mathcal{H}_c} | \Sigma^{(j)'} \rangle \propto p_{\mathrm{SA}} \left(\Sigma^{(j)}; \frac{\beta}{m} \right) \delta \left(\Sigma^{(j)}, \Sigma^{(j)'} \right), \tag{20}$$

$$\langle \Sigma^{(j)'} | e^{-\frac{\beta}{m} \mathcal{H}_q} | \Sigma^{(j+1)} \rangle \propto e^{s(\Sigma^{(j)'}, \Sigma^{(j+1)}) f(\beta, \Gamma)}, \tag{21}$$

since $\mathcal{H}_c$ is a diagonal matrix and the equation

$$(\sigma_i^x)^l = (\hat{1}_Q - \mathbb{E}_Q)^l = \frac{1}{Q} \left[(1 - Q)^l - 1 \right] \hat{1}_Q + (-1)^l \mathbb{E}_Q \tag{22}$$

is satisfied. Then we obtain

$$p_{\mathrm{QA}}(\Sigma; \beta, \Gamma) = \frac{1}{Z} \sum_{\Sigma^{(2)}} \cdots \sum_{\Sigma^{(m)}} \prod_{j=1}^{m} p_{\mathrm{SA}}(\Sigma^{(j)}; \frac{\beta}{m}) e^{s(\Sigma^{(j)}, \Sigma^{(j+1)}) f(\beta, \Gamma)}. \tag{23}$$

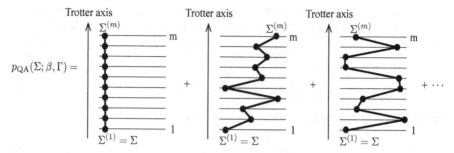

Fig. 2. Conceptual diagram of path integral. In order to calculate the probability of the state Σ of quantum system, we add an extra dimension which is called the Trotter axis. The probability $p_{QA}(\Sigma; \beta, \Gamma)$ can be calculated by taking sum of configurations of path depicted by the bold line.

By using path-integral representation, the probability of the state Σ of the d-dimensional quantum system can be represented by that of $(d+1)$-dimensional classical system approximately (see Fig.2).

Here it should be noted that the function $f(\beta, \Gamma)$ is a monotonic decreasing function of inverse temperature β and quantum field Γ (see Fig.3).

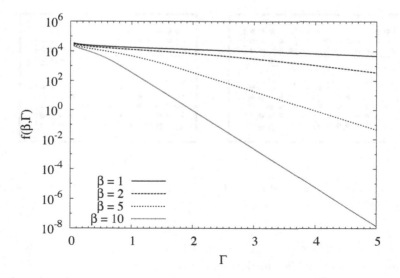

Fig. 3. $f(\beta, \Gamma)$ as a function of Γ for $N = 5000$, $Q = 30$, and $m = 50$.

In the quantum Monte Carlo simulation by using path-integral representation, we update the state according to the probability as follows:

$$p_{\text{QA-ST}}^{\text{update}}(\sigma_i^{(j)} = s|\Sigma^{(j)}\backslash\sigma_i^{(j)}, \Sigma^{(j-1)}, \Sigma^{(j+1)}; \beta, \Gamma) = \frac{P(s)}{\sum_{s'=1}^{Q} P(s')}, \quad (24)$$

$$P(s) = e^{\left\{-\frac{\beta}{m}E(\Sigma_s^{(j)})+[s(\Sigma^{(j-1)},\Sigma_s^{(j)})+s(\Sigma_s^{(j)},\Sigma^{(j+1)})]f(\beta,\Gamma)\right\}}, \quad (25)$$

where $\Sigma_s^{(j)}$ represents the state $\sigma_i^{(j)} = s$ given $\Sigma^{(j)}\backslash\sigma_i^{(j)}$.

5. Some Remarks

In the section 4, we introduced the quantum fluctuation and obtained the probability of the quantum state by the path-integral representation. Some remarks will be shown in this section.

5.1. *Labels of the clusters*

The value $s(\Sigma^{(j)}, \Sigma^{(j+1)})$ given by Eq. (18) expresses the correlation function between the state on the j-th Trotter layer and that on the $j + 1$-th Trotter layer. When the divisions are the same but the labels of the clusters are completely different as shown in Fig. 4, $s(\Sigma^{(j)}, \Sigma^{(j+1)})$ becomes zero.

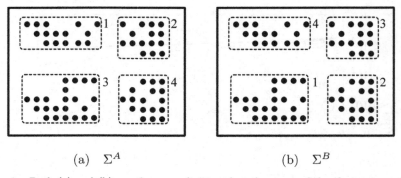

(a) Σ^A (b) Σ^B

Fig. 4. Both (a) and (b) are the same divisions but the name of the clusters are completely different.

When we decrease the quantum field Γ slow enough, not only division but also the labels of the clusters should become the same. If the state on the j-th layer is $\Sigma^{(j)} = \Sigma_A$ and that on the $j + 1$-th layer is $\Sigma^{(j+1)} = \Sigma_B$ such as Fig. 4 by accident, we cannot only gain a benefit but also are faced with a problem by introducing path-integral representation. This is similar situation with the domain wall problem in the ferromagnetic Ising model. In the ferromagnetic Ising model without external magnetic field, the ground state is that all spins are up or down. Suppose we consider the case that the spins in the left half are up and the spins in the right half are down. If we execute the standard Monte Carlo simulation, it is difficult to obtain the stable state (*i.e.* the ground state) from such an initial state for a short time. For clustering problems, the domain wall problem gets more seriously comparing with the standard ferromagnetic Ising model since the number of the same divisions is $Q!$. In order to avoid the domain wall problem, we introduce a new parameter "modified correlation function" and approximate the probability distribution. The definition of the modified correlation function $\tilde{s}(\Sigma^{(j)}, \Sigma^{(j+1)})$ is

$$\tilde{s}(\Sigma^{(j)}, \Sigma^{(j+1)}) := \frac{1}{N} \sum_{c=1}^{k} \max_{c'=1,\cdots,k} [Y(\Sigma^{(j)})Y^T(\Sigma^{(j+1)})]_{c,c'}, \quad (26)$$

$$Y(\Sigma^{(j)}) := (\sigma_1^{(j)}, \sigma_2^{(j)}, \cdots, \sigma_N^{(j)}), \quad (27)$$

where $Y(\Sigma^{(j)})$ denotes $N \times Q$ matrix. For $Q = 3$ and $N = 2$, $Y(\Sigma_3)$, where Σ_3 was defined in Table 1, is given as

$$Y(\Sigma_3) = \begin{pmatrix} 0 & 1 \\ 0 & 0 \\ 1 & 0 \end{pmatrix}. \quad (28)$$

It should be noted that the correlation function $s(\Sigma^{(j)}, \Sigma^{(j+1)})$ can be expressed in a similar way such as

$$s(\Sigma^{(j)}, \Sigma^{(j+1)}) = \frac{1}{N} \mathrm{Tr}[Y(\Sigma^{(j)})Y^T(\Sigma^{(j+1)})]. \quad (29)$$

Let us show some properties of the modified correlation function comparing $s(\Sigma^{(j)}, \Sigma^{(j+1)})$ in the next section. Suppose we consider the case for $Q = 3$ and $N = 7$ in the section 5.1.1 and 5.1.2.

5.1.1. *Example A*

We consider a case that the states on the j-th layer and that on the $j+1$-th layer are shown in Table 2.

Here, $Y(\Sigma^{(j)})$ and $Y(\Sigma^{(j+1)})$ are given as

$$Y(\Sigma^{(j)}) = \begin{pmatrix} 1 & 1 & 1 & 0 & 0 & 0 & 0 \\ 0 & 0 & 0 & 1 & 1 & 0 & 0 \\ 0 & 0 & 0 & 0 & 0 & 1 & 1 \end{pmatrix}, \quad Y(\Sigma^{(j+1)}) = \begin{pmatrix} 0 & 0 & 0 & 0 & 1 & 1 & 0 \\ 0 & 0 & 1 & 0 & 0 & 0 & 1 \\ 1 & 1 & 0 & 1 & 0 & 0 & 0 \end{pmatrix}. \quad (30)$$

Table 2. Example A.

	σ_1	σ_2	σ_3	σ_4	σ_5	σ_6	σ_7
$\Sigma^{(j)}$	1	1	1	2	2	3	3
$\Sigma^{(j+1)}$	3	3	2	3	1	1	2

Then, the modified correlation function is calculated as

$$\tilde{s}(\Sigma^{(j)}, \Sigma^{(j+1)}) = \frac{1}{7} \sum_{c=1}^{3} \max_{c'=1,\cdots,3} \begin{pmatrix} 0 & 1 & 2 \\ 1 & 0 & 1 \\ 1 & 1 & 0 \end{pmatrix}_{c,c'} = \frac{4}{7}. \tag{31}$$

Note that $s(\Sigma^{(j)}, \Sigma^{(j+1)}) = 0$ by the definition. Next we fix the labels of $\Sigma^{(j)}$ and rename the labels of $\Sigma^{(j+1)}$. The values of $s(\Sigma^{(j)}, \mathcal{P}_\pi \Sigma^{(j+1)})$ are shown in Table 3, where $\mathcal{P}_\pi$ ($\pi = 1, \cdots, Q!$) denotes label permutation operator.

Table 3. The value of correlation function by applying label permutation operator $\mathcal{P}_\pi$ for example A.

	σ_1	σ_2	σ_3	σ_4	σ_5	σ_6	σ_7	$s(\Sigma^{(j)}, \mathcal{P}_\pi \Sigma^{(j+1)})$
$\mathcal{P}_1 \Sigma^{(j+1)}$	3	3	2	3	1	1	2	0
$\mathcal{P}_2 \Sigma^{(j+1)}$	3	3	1	3	2	2	1	2/7
$\mathcal{P}_3 \Sigma^{(j+1)}$	1	1	3	1	2	2	3	4/7
$\mathcal{P}_4 \Sigma^{(j+1)}$	1	1	2	1	3	3	2	3/7
$\mathcal{P}_5 \Sigma^{(j+1)}$	2	2	3	2	1	1	3	2/7
$\mathcal{P}_6 \Sigma^{(j+1)}$	2	2	1	2	3	3	1	3/7

In this case, the maximum value of $s(\Sigma^{(j)}, \mathcal{P}_\pi \Sigma^{(j+1)})$ is the same as the value of modified correlation function $\tilde{s}(\Sigma^{(j)}, \Sigma^{(j+1)})$.

5.1.2. Example B

We also consider another case that the states on the j-th layer and that on the $j + 1$-th layer are shown in Table 4 as an another example.

Table 4. Example B.

	σ_1	σ_2	σ_3	σ_4	σ_5	σ_6	σ_7
$\Sigma^{(j)}$	1	2	2	2	2	3	3
$\Sigma^{(j+1)}$	2	1	1	1	3	1	1

Here, $Y(\Sigma^{(j)})$ and $Y(\Sigma^{(j+1)})$ are given as

$$Y(\Sigma^{(j)}) = \begin{pmatrix} 1\,0\,0\,0\,0\,0\,0 \\ 0\,1\,1\,1\,1\,0\,0 \\ 0\,0\,0\,0\,0\,1\,1 \end{pmatrix}, \quad Y(\Sigma^{(j+1)}) = \begin{pmatrix} 0\,1\,1\,1\,0\,1\,1 \\ 1\,0\,0\,0\,0\,0\,0 \\ 0\,0\,0\,0\,1\,0\,0 \end{pmatrix}. \quad (32)$$

Then, the modified correlation function is calculated as

$$\tilde{s}(\Sigma^{(j)}, \Sigma^{(j+1)}) = \frac{1}{7} \sum_{c=1}^{3} \max_{c'=1,\cdots,3} \begin{pmatrix} 0\,1\,0 \\ 3\,0\,1 \\ 2\,1\,1 \end{pmatrix}_{c,c'} = \frac{6}{7}. \quad (33)$$

Note that $s(\Sigma^{(j)}, \Sigma^{(j+1)}) = 0$ by the definition. As in the section 5.1.1, when we fix the labels of $\Sigma^{(j)}$ and rename the labels of $\Sigma^{(j+1)}$, the values of $s(\Sigma^{(j)}, \mathcal{P}_\pi \Sigma^{(j+1)})$ are shown in Table 5.

Table 5.   The value of correlation function by applying label permutation operator $\mathcal{P}_\pi$ for example B.

	σ_1	σ_2	σ_3	σ_4	σ_5	σ_6	σ_7	$s(\Sigma^{(j)}, \mathcal{P}_\pi \Sigma^{(j+1)})$
$\mathcal{P}_1 \Sigma^{(j+1)}$	2	1	1	1	3	1	1	0
$\mathcal{P}_2 \Sigma^{(j+1)}$	3	1	1	1	2	1	1	1/7
$\mathcal{P}_3 \Sigma^{(j+1)}$	1	2	2	2	3	2	2	4/7
$\mathcal{P}_4 \Sigma^{(j+1)}$	3	2	2	2	1	2	2	3/7
$\mathcal{P}_5 \Sigma^{(j+1)}$	1	3	3	3	2	3	3	4/7
$\mathcal{P}_6 \Sigma^{(j+1)}$	2	3	3	3	1	3	3	2/7

In this case, the maximum value of $s(\Sigma^{(j)}, \mathcal{P}_\pi \Sigma^{(j+1)})$ is not the same as the modified correlation function $\tilde{s}(\Sigma^{(j)}, \Sigma^{(j+1)})$. Although the modified correlation function expresses some kinds of similarity between the states $\Sigma^{(j)}$ and $\Sigma^{(j+1)}$, it is not necessary commutative $i.e.$ $\tilde{s}(\Sigma^{(j)}, \Sigma^{(j+1)}) \neq \tilde{s}(\Sigma^{(j+1)}, \Sigma^{(j)})$. In the case of example A,

$$\tilde{s}(\Sigma^{(j)}, \Sigma^{(j+1)}) = \tilde{s}(\Sigma^{(j+1)}, \Sigma^{(j)}) = \frac{4}{7}. \quad (34)$$

On the other hand, in the case of example B,

$$\tilde{s}(\Sigma^{(j)}, \Sigma^{(j+1)}) = \frac{6}{7}, \quad \tilde{s}(\Sigma^{(j+1)}, \Sigma^{(j)}) = \frac{5}{7}. \quad (35)$$

Here we summarize the properties of the modified correlation function:

- The modified correlation function does not necessary commute.

$$\tilde{s}(\Sigma^{(j)}, \Sigma^{(j+1)}) \neq \tilde{s}(\Sigma^{(j+1)}, \Sigma^{(j)}). \quad (36)$$

- For any $\mathcal{P}_\pi$ $(\pi = 1, \cdots, Q!)$, the inequality

$$0 \leq s(\Sigma^{(j)}, \mathcal{P}_\pi \Sigma^{(j+1)}) = \mathrm{Tr}\left[Y(\Sigma^{(j)})Y(\mathcal{P}_\pi \Sigma^{(j+1)})^T\right] \leq \tilde{s}(\Sigma^{(j)}, \Sigma^{(j+1)}) \leq 1$$

  is satisfied.

Although the modified correlation function is not necessary the same as the maximum value of $s(\Sigma^{(j)}, \mathcal{P}_\pi \Sigma^{(j+1)})$, we use this value in our proposed algorithm. This is because our aim is to find not so bad solution of given problem as mentioned before.

5.2. Speed-up by using modified correlation function

When we adopt the quantum Monte Carlo simulation by the path-integral representation, it is enough to update the state according to Eq. (24) in principle. In the previous section, we introduced the new parameter "modified correlation function" to avoid the so-called domain wall problem. Instead of Eq. (24), we adopt the following the transition probability based on modified correlation function:

$$p_{\mathrm{QA-ST+modify}}^{\mathrm{update}}(\sigma_i^{(j)} = s|\Sigma^{(j)}\backslash\sigma_i^{(j)}, \Sigma^{(j-1)}, \Sigma^{(j+1)}; \beta, \Gamma)$$

$$= \frac{\exp\left[-\frac{\beta}{m}E(\Sigma_s^{(j)}) + \tilde{S}(\Sigma^{(j-1)}, \Sigma_s^{(j)}, \Sigma^{(j+1)})f(\beta, \Gamma)\right]}{\sum_{t=1}^{Q} \exp\left[-\frac{\beta}{m}E(\Sigma_t^{(j)}) + \tilde{S}(\Sigma^{(j-1)}, \Sigma_t^{(j)}, \Sigma^{(j+1)})f(\beta, \Gamma)\right]}, \qquad (37)$$

$$\tilde{S}(\Sigma^{(j-1)}, \Sigma_s^{(j)}, \Sigma^{(j+1)}) := \tilde{s}(\Sigma^{(j-1)}, \Sigma_s^{(j)}) + \tilde{s}(\Sigma_s^{(j)}, \Sigma^{(j+1)}). \qquad (38)$$

Since $\tilde{s}(\Sigma^{(j-1)}, \Sigma^{(j)})$ is not necessary commutative, there are four possibilities of the definition of $\tilde{S}(\Sigma^{(j-1)}, \Sigma^{(j)}, \Sigma^{(j+1)})$ as follows:

$$\tilde{S}(\Sigma^{(j-1)}, \Sigma_s^{(j)}, \Sigma^{(j+1)}) = \tilde{s}(\Sigma^{(j-1)}, \Sigma_s^{(j)}) + \tilde{s}(\Sigma_s^{(j)}, \Sigma^{(j+1)}), \qquad (39)$$

$$\tilde{S}(\Sigma^{(j-1)}, \Sigma_s^{(j)}, \Sigma^{(j+1)}) = \tilde{s}(\Sigma^{(j-1)}, \Sigma_s^{(j)}) + \tilde{s}(\Sigma^{(j+1)}, \Sigma_s^{(j)}), \qquad (40)$$

$$\tilde{S}(\Sigma^{(j-1)}, \Sigma_s^{(j)}, \Sigma^{(j+1)}) = \tilde{s}(\Sigma_s^{(j)}, \Sigma^{(j-1)}) + \tilde{s}(\Sigma_s^{(j)}, \Sigma^{(j+1)}), \qquad (41)$$

$$\tilde{S}(\Sigma^{(j-1)}, \Sigma_s^{(j)}, \Sigma^{(j+1)}) = \tilde{s}(\Sigma_s^{(j)}, \Sigma^{(j-1)}) + \tilde{s}(\Sigma^{(j+1)}, \Sigma_s^{(j)}). \qquad (42)$$

For small systems we confirm that when we adopt Eq. (39) as the definition of $\tilde{S}(\Sigma^{(j-1)}, \Sigma_s^{(j)}, \Sigma^{(j+1)})$, we can obtain better solution than the other choices. Then, we adopt the relation given by Eq. (39) as the definition of $\tilde{S}(\Sigma^{(j-1)}, \Sigma_s^{(j)}, \Sigma^{(j+1)})$.

5.3. *Simultaneous control of thermal fluctuation and quantum fluctuation*

In our algorithm given by Eq. (37), thermal fluctuation β/m and quantum fluctuation $f(\beta, \Gamma)$ coexist. Then, it is expected that we can obtain the better solution than simulated annealing by controlling the thermal fluctuation and quantum fluctuation simultaneously with ingenuity.

In order to investigate how to control both of the thermal fluctuation and quantum fluctuation, we first consider limiting cases. When the temperature is very low than the quantum field, *i.e.* $\beta/m \gg f(\beta, \Gamma)$, the probability distributions in each layer obey Eq. (13). On the other hand, if the quantum field is very weak, *i.e.* $\beta/m \ll f(\beta, \Gamma)$, the state in each layers becomes the same state such as $\Sigma = \Sigma^{(j)}$ for all j. By calculating small systems, we find how to control the thermal fluctuation and quantum fluctuation simultaneously. The best condition is as follows: (i) At the first step, to find the metastable state on each layers, we set the temperature $\beta/m \gg f(\beta, \Gamma)$. (ii)After that, $f(\beta, \Gamma)$ overstrides thermal fluctuation. The curve f^* in Fig. 5 shows a conceptual diagram for our proposed schedule.

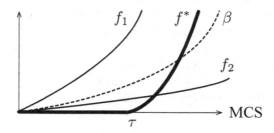

Fig. 5. The dotted curve denotes the schedule of inverse temperature. f_1 is too rapid for decreasing the quantum field. On the other hand, f_2 is too slow for decreasing the quantum field. f^* denotes the best schedule.

The dotted curve in Fig. 5 indicates the schedule of cooling temperature which corresponds to the simulated annealing. The curve depicted f^* in Fig. 5 is the best schedule for obtaining better solution than the simulated annealing. In the schedule depicted f_1 in Fig. 5, quantum mixing effect does not make sense. In the schedule depicted f_2 in Fig. 5, on the other hand, the states on each layers behave independently. The quantum fluctuation effect is strong than the thermal fluctuation effect in the schedule depicted f_2. It is essentially the same as the simulated annealing.

Here we assume the scheduling functions of temperature and quantum field as follows:

$$\beta(t) = \beta_0 r_\beta^t, \tag{43}$$

$$\Gamma(t) = \infty, \quad (t < \tau), \qquad \Gamma(t) = \Gamma_0 \exp(-r_\Gamma^{t-\tau}) \quad (t \geq \tau), \tag{44}$$

where τ corresponds to the time $\beta(\tau) = m$. When $Q\beta\Gamma/m \ll 1$, the interaction along the Trotter axis $f(\beta, \Gamma)$ given by Eq. (19) is approximately given as

$$f(\beta, \Gamma) \sim -N \log(\frac{\beta\Gamma}{m}) = Nr_\Gamma^t - N \log(\frac{\beta\Gamma_0}{m}). \tag{45}$$

From this equation, it is enough to set large enough Γ_0 and $r_\beta < r_\Gamma$ in order to prepare the schedule f^*. In other words, independent simulated annealing is performed until τ and after that we decrease quantum field to obtain the better solution than the conventional simulated annealing.

6. Results

We perform numerical experiment for the following three problems by proposed quantum annealing and simulated annealing for comparison. We prepare the number of replicas $m = 50$ in the whole experiments. The initial inverse temperature and the initial quantum field are set to be $\beta_0 = 0.2m$ and $\Gamma_0 = e^{1/2}$, respectively. Here the ratio of changing temperature is set to be $r_\beta = 1.05$. In the simulated annealing, we prepare independent 55 samples and use the same initial temperature and the ratio of changing temperature as the case of quantum annealing. We study three problems as follows:

(a) Evaluation of mixture of Gaussian by using MNIST data[40]
     $(Q = 30, N = 5000)$
(b) Evaluation of latent Dirichlet allocation[41] by using Reuters data[42]
     $(Q = 20, N = 2000)$
(c) Evaluation of latent Dirichlet allocation by using NIPS corpus[43]
     $(Q = 20, N = 1000)$

In all of the experiments, we fix the ratio of changing the inverse temperature $r_\beta = 1.05$.

Figure 6 denotes the time development of the minimum energy in all of the layers for each problems. The circles in Fig. 6 represent results obtained by simulated annealing for comparison. The triangles, asterisks, and squares denote obtained results by proposed quantum annealing for $r_\Gamma = 1.02$, $r_\Gamma = 1.10$, and $r_\Gamma = 1.20$, respectively. The schedule for $r_\Gamma = 1.02$ corresponds

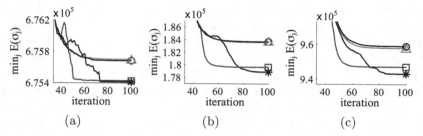

Fig. 6. Time development of the minimum energy in all of the layers. The circles represent results obtained by simulated annealing for comparison. The triangles, asterisks, and squares denote obtained results by proposed quantum annealing for $r_\Gamma = 1.02$, $r_\Gamma = 1.10$, and $r_\Gamma = 1.20$, respectively. (a) Evaluation of mixture of Gaussian by using MNIST data ($Q = 30, N = 5000$). (b) Evaluation of latent Dirichlet allocation by using Reuters data ($Q = 20, N = 2000$). (c) Evaluation of latent Dirichlet allocation by using NIPS corpus ($Q = 20, N = 1000$).

to schedule depicted f_2 in Fig. 5 whereas the schedules for $r_\Gamma = 1.10$ and $r_\Gamma = 1.20$ correspond to schedule depicted f^* in Fig. 5. From Fig. 6 it is clear that we can obtain the better solution than simulated annealing if we adopt the schedule f^*.

7. Conclusion

In this paper, we developed hybrid type of the quantum annealing for clustering problems. When we apply the standard quantum annealing for these problems, we face on the difficulty of the domain wall problem. Since the number of the same division is $Q!$, the domain wall problem gets more serious for large Q. To avoid such a problem, we introduced new parameter "modified correlation function" instead of the standard correlation function. We investigated the best schedule of changing the thermal fluctuation and the quantum fluctuation simultaneously. We first apply strong quantum field and then we obtain the metastable states. At the second step, we can obtain the better solution than the simulated annealing by decreasing the quantum field. Actually, we succeeded to obtain the better solution for clustering problems than the simulated annealing. We expect that the proposed schedule of changing the thermal fluctuation and the quantum fluctuation is generally efficient for other type of problems. However, it is an open problem when to use the quantum annealing. To solve this problem, we should study the efficiency of quantum annealing for the problems where the *difficulty* of the problem can be controlled.

The authors are grateful to Naoki Kawashima, Seiji Miyashita, Hiroshi

Nakagawa, Daiji Suzuki, and member of T-PRIMAL for their valuable comments. S.T. is partly supported by Grant-in-Aid for Young Scientists Start-up (21840021) from the JSPS, MEXT Grant-in-Aid for Scientific Research (B) (22340111), and the "Open Research Center" Project for Private Universities: matching fund subsidy from MEXT. R.T. is partly supported by Global COE Program "the Physical Sciences Frontier", MEXT, Japan. The computation in the present work was performed on computers at the Supercomputer Center, Institute for Solid State Physics and University of Tokyo and at Taisuke Sato's group, Tokyo Institute of Technology.

Appendix A. Implementation Method of New Kind of Fluctuation

In this paper, we studied a quantum effect on clustering problems which are expressed by the Potts model. Relationship between a phase transition and a performance of quantum annealing is very important, although we did not mention it in this paper. Many researchers have studied the order of the phase transition of a given problem and concluded whether the quantum annealing is efficient or not for the problems.[8,10,44–46] It is one of main topics on quantum annealing.

Suppose there is a transition point at halfway of a control parameter such as magnetic field. If a first-order phase transition occurs, the quantum adiabatic computation, in principle, does not obtain a good solution because of a level-crossing problem. There is a trivial example of level-crossing problem: the ferromagnetic Ising model with longitudinal magnetic field h^z. In this model, level crossing occurs at $h^z = 0$. The initial state is set to be a ground state of negative h^z. When we sweep h^z from negative to positive, the state cannot follow the adiabatic limit of the state at all. If a second-order phase transition occurs, on the other hand, the growth of correlation length does not follow for finite speed of sweeping of control parameter. The equilibrium value of correlation length diverges at the second-order phase transition point. In practice, the quantum annealing does not succeed to obtain the best solution for systems in which the second-order phase transition occurs. If there is no phase transition, the quantum annealing is expected to find the best solution.

Our purposes are to control the order of phase transition and, what is more, to erase the phase transition by adding some kind of fluctuation. Recently, we constructed a model in which the order of the phase transition can be changed by controlling a fluctuation.[47–49] This model is called the Potts model with invisible states. We found that a first-order phase transition is driven by the effect of invisible states (invisible fluctuation) in the ferromagnetic Potts model with invisible states. Although the invisible fluctuation itself seems to be getting worse for quantum annealing from the above discussion, the invisible fluctuation changes the order of the phase transition without changing an essence of problems. We expect that there are fluctuations which wipe a phase transition. Thus, to introduce new kinds of fluctuation is important for optimization problems. In this section, we introduce the concept of invisible fluctuation.

We first consider the standard Potts model. The Hamiltonian of this model is given by

$$\mathcal{H}_{\text{standard}} = \sum_{\langle i,j \rangle \in E(G)} J_{ij} \delta_{\sigma_i, \sigma_j}, \qquad \sigma_i = 1, \cdots, Q, \qquad (A.1)$$

where $E(G)$ denotes the set of edges of given graph G. Eq. (A.1) is called the Q-state Potts model. Here we assume Q is a natural number. Suppose we consider the ferromagnetic case i.e. $J_{ij} = -J$ for all $\langle i, j \rangle \in E(G)$. A second-order phase transition occurs when $Q \leq 4$ whereas a first-order phase transition occurs when $Q > 4$ on two-dimensional lattice. It is interesting that the order of the phase transition of the standard ferromagnetic Potts model can be changed by the number of states Q. The ground state of this model is that all of the spins have the same value. The number of ground states is Q. Then, the phase transition accompanies spontaneous Q-fold symmetry breaking. The standard Potts model has been regarded as the standard model not only in statistical physics but also in wide area of science.

We consider the Hamiltonian of the Potts model with invisible states as follows:

$$\mathcal{H}_{\text{inv}} = \sum_{\langle i,j \rangle \in E(G)} J_{ij} \delta_{\sigma_i, \sigma_j} \sum_{\alpha=1}^{Q} \delta_{\sigma_i, \alpha}, \qquad \sigma_i = 1, \cdots, Q + R. \qquad (A.2)$$

This model is called the (Q,R)-state Potts model.[47–49] Suppose we consider the case for $J_{ij} = -J$ for all $\langle i, j \rangle \in E(G)$ for simplicity as the previous example. If and only if $1 \leq \sigma_i = \sigma_j \leq Q$, interaction $-J$ works. Thus, the number of ground states is Q. Note that if $R = 0$, this model is equivalent to the standard ferromagnetic Potts model. Hereafter we call the states where $1 \leq \sigma_i \leq Q$ "colored states" whereas the states where $Q + 1 \leq \sigma_i \leq Q + R$ "invisible states".

Here we consider two spin system. The number of excited states of the standard ferromagnetic Potts model given by Eq. (A.1) is $Q^2 - Q$. On the other hand, the number of excited states of the (Q,R)-state Potts model given by Eq. (A.2) is $Q^2 - Q + 2QR + R^2$. The (Q,R)-state Potts model does not change the number of degeneracy of the ground states. However the number of excited states are different. As the number of sites N increases, since density of state changes, it is expected that nature of phase transition changes.

The order parameter of the (Q,R)-state Potts model is defined as

$$\mathbf{m} = \frac{1}{N} \sum_{i=1}^{N} \mathbf{e}^{\sigma_i}, \tag{A.3}$$

where $\mathbf{e}^\alpha$ $(\alpha = 1, \cdots, Q)$ represents Q unit vectors pointing in the Q symmetric direction of a hypertetrahedron in $Q - 1$ dimensions. It should be noted that the Kronecker's delta can be represented by using $\mathbf{e}^\alpha$ as follows:

$$\delta_{\alpha,\beta} = \frac{1 + (Q - 1)\mathbf{e}^\alpha \cdot \mathbf{e}^\beta}{Q}. \tag{A.4}$$

The definition of order parameter is the same as that of the standard ferromagnetic Q-state Potts model. The phase transition accompanies Q-fold symmetry if a phase transition takes place in this model. We investigated this model by mean-field analysis and Monte Carlo simulation.[47-49] In these papers, we concluded that the invisible states drive the first-order phase transition and a phase transition with Q-fold symmetry breaking occurs at finite temperature.

Before concluding this section, we discuss why a first-order phase transition is driven by the invisible fluctuation. The Hamiltonian given in Eq. (A.2) can be transformed exactly by comparing the partition function as follows:

$$\mathcal{H}_{\text{inv}}^{\text{eff}} = \sum_{\langle i,j \rangle \in E(G)} J_{ij} \delta_{\tau_i,\tau_j} \sum_{\alpha=1}^{Q} \delta_{\tau_i,\alpha} - T \log r \sum_{i=1}^{N} \delta_{\tau_i,0}, \tag{A.5}$$

$$\tau_i = 0, 1, \cdots, Q, \tag{A.6}$$

where T represents a temperature. Here we rename the label of the invisible states from $Q + 1 \le \sigma_i \le Q + R$ to $\tau_i = 0$. The second term means chemical potential of the invisible states. The (Q,R)-state Potts model can be mapped onto the annealed diluted Potts model whose chemical potential depends on temperature linearly. As we change temperature, the chemical potential is varied. This concept is similar with our hybrid quantum annealing method. It should be noted that temperature-dependency of the chemical potential comes from the number of invisible states, in other words, entropy of the invisible states.

As mentioned above, the invisible fluctuation itself is inefficient for optimization problem. However the quantum annealing expects to be a powerful method by adding a new fluctuation – quantum fluctuation. In a similar way, it is possible that there is a "good" fluctuation for optimization problems. In this section, we have considered the effect of invisible fluctuation. The invisible fluctuation is one of "entropic fluctuation". The order of phase transition is decided by the density of states. Then, such a entropic fluctuation is expected to wipe a phase transition. We believe that an entropic fluctuation which is constructed as the invisible fluctuation makes some advantages for optimization problems.

References

1. S. Kirkpatrick, C. D. Gelatt, and M. P. Vecchi, *Science* **220**, 671 (1983).
2. S. Kirkpatrick, *J. Stat. Phys.* **34**, 975 (1984).
3. S. Geman and D. Geman, *IEEE Trans. Pattern Anal. Mach. Intell.* **PAMI-6**, 721 (1984).
4. T. Kadowaki and H. Nishimori, *Phys. Rev. E* **58**, 5355 (1998).
5. E. Farhi, J. Goldstone, S. Gutmann, J. Lapan, A. Lundgren, and D. Preda, *Science* **292** 472 (2001).
6. G.E. Santoro, R. Martoncaronák, E. Tosatti, and R. Car, *Science* **295** 2527 (2002).
7. T. Kadowaki, arXiv:quant-ph/0205020.
8. A. Das and B.K. Charkrabarti, *Quantum Annealing And Related Optimization Methods (Lecture Notes in Physics)* (Springer-Verlag, 2005).
9. A. Das and B.K. Charkrabarti, *Rev. Mod. Phys.* **80** 1061 (2008).
10. A.K. Chandra, A. Das, B.K. Charkrabarti, *Quantum Quenching, Annealing and Computation (Lecture Notes in Physics)* (Springer Berlin Heidelberg, 2010).
11. M. Ohzeki and H. Nishimori, *J. Comp. and Theor. Nanoscience* **8** 963 (2011).
12. K. Kurihara, S. Tanaka, and S. Miyashita, *Proceedings of the 25th Conference on Uncertainty in Artificial Intelligence* (2009).
13. T. Nakamura, *Phys. Rev. Lett.* **101** 210602 (2008).
14. S. Morita, S. Suzuki, and T. Nakamura, *Phys. Rev. E* **79** 065701 (2009).
15. S. Suzuki and M. Okada, *Interdisciplinary Information Sciences* **13** 49 (2007).
16. J. Rodríguez-Laguna, *J. Stat. Mech.* P05008 (2007).
17. K. Tanaka, T. Horiguchi, *Interdisciplinary Information Sciences* **8** 33 (2002).
18. I. Sato, K. Kurihara, S. Tanaka, H. Nakagawa, and S. Miyashita, *Proceedings of the 25th Conference on Uncertainty in Artificial Intelligence* (2009).
19. E. Jané, G. Vidal, W. Dür, P. Zollar, and J.I. Cirac, *Quantum Information and Computation* **3** 15 (2003).
20. D. Porras and J.I. Cirac, *Phys. Rev. Lett.* **92** 207901 (2004).

21. M. Lewenstein, A. Sanpera, V. Ahufinger, B. Damski, and A. Sen, *Adv. in Phys.* **56** 243 (2007).

22. A. Friedenauer, H. Schmitz, J.T. Glueckert, D. Porras, and T. Schaetz, *Nat. Phys.* **4** 757 (2008).

23. R. Harris, A.J. Berkley, J. Johansson, M.W. Johnson, T. Lanting, P. Bunyk, E. Tolkacheva, E. Ladizinsky, B. Bumble, A. Fung, A. Kaul, A. Kleinsasser, S. Han, arXiv:0903.3906 (2009).

24. A.J. Berkley, M.W. Johnson, P. Bunyk, R. Harris, J. Johansson, T. Lanting, E. Ladizinsky, E. Tolkacheva, M.H.S. Amin, and G. Rose, *Supercond. Sci. Technol.* **23** 105014 (2010).

25. R. Harris, J. Johansson, A.J. Berkley, M.W. Johnson, T. Lanting, Siyuan Han, P. Bunyk, E. Ladizinsky, T. Oh, I. Perminov, E. Tolkacheva, S. Uchaikin, E.M. Chapple, C. Enderud, C. Rich, M. Thom, J. Wang, B. Wilson, and G. Rose, *Phys. Rev. B* **81** 134510 (2010).

26. M.H.S. Amin, N.G. Dickson, and P. Smith, arXiv:1103.1904 (2011).

27. S. Morita and H. Nishimori, *J. Phys. A* **39** 13903 (2006).

28. S. Morita and H. Nishimori, *J. Phys. Soc. Jpn.* **76** 064002 (2007).

29. S. Morita and H. Nishimori, *J. Math. Phys.* **49** 125210 (2008).

30. S. Tanaka and S. Miyashita, *J. Magn. Magn. Mater.* **310** e468 (2007).

31. Y. Matsuda, H. Nishimori, and H.G. Katzgraber, *New. J. Phys.* **11** 073021 (2009).

32. S. Miyashita, S. Tanaka, H.de Raedt, and B. Barbara, *J. Phys.: Conference Series* **143** 012005 (2009).

33. S. Tanaka, M. Hirano, and S. Miyashita, *Lecture Note in Physics* **802** 215 (2010).

34. S. Tanaka and S. Miyashita, *Phys. Rev. E* **81** 051138 (2010).

35. S. Tanaka, M. Hirano, and S. Miyashita, *Physica E* **43** 766 (2010).

36. M. Ohzeki, *Phys. Rev. Lett.* **105** 050401 (2010).

37. K. Hukushima and K. Nemoto, *J. Phys. Soc. Jpn.* **65** 1604 (1996).

38. H.F. Trotter, *Proc. Amer. Math. Soc.* **56** 1454 (1959).

39. M. Suzuki, *Prog. Theor. Phys.* **56** 1454 (1976).

40. http://yann.lecun.com/exdb/mnist/

41. D.M. Blei, A.Y. Ng, and M.I. Jordan, *The Journal of Machine Learning Research* **3** 993 (2003).

42. http://www.cs.nyu.edu/~roweis/data.html

43. http://books.nips.cc/

44. T. Caneva, R. Fazio, and G.E. Santoro, *Phys. Rev. B* **76**, 144427 (2007).

45. S. Suzuki, *J. Stat. Mech.* P03032 (2009).

46. A.P. Young, S. Kynsh, and V.N. Smelyanskiy *Phys. Rev. Lett.* **104** 020502 (2010).

47. R. Tamura, S. Tanaka, and N. Kawashima, *Prog. Theor. Phys.* **124**, 381 (2010).

48. S. Tanaka and R. Tamura, arXiv:1012.4254. (2010).

49. S. Tanaka, R. Tamura, and N. Kawashima, arXiv:1102.5475. (2011).

THEORY OF MACROSCOPIC QUANTUM TUNNELING IN TWO-GAP SUPERCONDUCTORS: CORRECTIONS FROM LEGGETT'S MODES

YUKIHIRO OTA[*,‡,¶], MASAHIKO MACHIDA[*,‡,§] and TOMIO KOYAMA[†,§]

CCSE, Japan Atomic Energy Agency,
6-9-3 Higashi-Ueno Taito-ku, Tokyo 110-0015, Japan
†*Institute for Materials Research, Tohoku University,*
2-1-1 Katahira Aoba-ku, Sendai 980-8577, Japan
‡*CREST(JST), 4-1-8 Honcho, Kawaguchi, Saitama 332-0012, Japan*
§*JST-TRIP, 5 Sambancho Chiyoda-ku, Tokyo 102-0075, Japan*
¶*E-mail: yota@alice.math.kindai.ac.jp*

We theoretically examine macroscopic quantum tunneling (MQT) in a Josephson junction formed by a conventional single-gap Bardeen-Cooper-Schrieffer superconductor and a multigap one such as magnesium diboride and iron-pnictide compounds. We clarify the quantum dynamics of gauge-invariant phase differences and construct a theory for the MQT in the multigap Josephson junctions. The theory predicts that the MQT escape rate is drastically enhanced when the frequency of Leggett's mode is less than the Josephson-plasma frequency.

Keywords: Macroscopic Quantum Tunneling, Multiband Superconductivity, Leggett's Modes

1. Introduction

Macroscopic quantum tunneling (MQT)[1-3] is occurrence of quantum coherence in macroscopic level and has been studied in various physical area such as condensed matters, nuclei, cosmology, etc. Intriguing realization of MQT is a switching event in Josephson junctions at low temperatures.[4,5] The MQT in Josephson junctions has been intensively studied since it is promising for applications to a Josephson phase qubit.[6-8]

Josephson effects strongly depends on Cooper pair's properties in the superconducting electrodes. Discovery of new superconducting materials may require theoretical extension of Josephson effects and open a novel route of the application. Since the discovery of novel high-T_c iron-based superconductors,[9] a study on multiband superconductivity has been revived. One of the key features is the presence of multiple superconducting gaps below the superconduciting transition temperature. Josephson effects are considerably alterd in multiband superconductors such as the iron-based superconductors and the magnesium diboride (MgB_2)[10] because the appearance of multiple tunneing channels ascribed to the the multiband brings about multiple degrees of freedom (i.e., multiple gauge-invariant phase differences) in Josephson junctions.[11] Many researchers have intensively examined nontrivial (classical) Josephson effects associated with the multiple tunneling channels,[11-19] while few studies about the MQT exists.[20]

The purpose of this paper is to develop a theory of the MQT in Josephson junctions with multiple tunneling channels. We show that the MQT escape rate is significantly enhanced by taking into account of zero point motion with respect to relative phases between superconducting phases, focusing on a simple but essential model for the Josephson junction with multiple tunneling channels, a hetero Josephson junction between one- and two-gap superconductors. In particular, such enhancement is predominant when Leggett's mode [21-23] frequency locates at a low-energy regime compared to the Josephson-plasma frequency. We also mention that the present system has a new characteristic scale, the ratio of the Josephson energy E_J to the inter-band Josephson energy E_{in}, in addition to a typical characteristic quantity in Josephson junctions E_J/E_C, in which E_C is the charge energy. Then, we successfully predict a situation that the MQT escape rate becomes large by tuning E_J/E_{in} for fixed E_J/E_C.

The present theory reveals the presence of rich phenomena in an unexplored field, i.e., the MQT in the multiband superconductivity. A key quantity is a relative phase between the superconducting phases, which induces drastic enhancement of the MQT escape rate. The theory of the MQT is applicable to various physical issues from a condensed matter system to cosmology. Hence, the present contribution has a great impact on all of the quantum tunneling phenomena.

2. Variety of Superconductivity

A standard Josephson qubit is made of a conventional low-T_c Bardeen-Cooper-Schrieffer (BCS) type superconductor,[24] such as Al. Such a kind of the system has a lot of advantages for physical realization of quantum information processing and quantum engineering. A sophisticate fabrication technique of electric circuits leads to a flexible design of a quantum system in which a long-lived quantum coherence manifests with a macroscopic scale. In particular, a resent development of the circuit QED [25] allows us to create a large scale quantum network system. It is worth while mentioning here that the BCS type superconductor is not a unique material occurring superconductivity and Josephson effects. In this section, we briefly explain three types of superconductivity other than the BCS type one and their application potential to quantum information processing. In particular, we focus on High-T_c copper oxide, chiral superconductivity, and multiband superconductivity. A comprehensive review about unconventional superconductivity including multiband or multicomponent is given in Ref. 26.

2.1. High-T_c copper oxide and intrinsic Josephson junction arrays

High-T_c copper oxide, such as $Bi_2Sr_2CaCu_2O_8$, which has a much higher superconducting transition temperature than of the standard BCS type superconductors, is a notable superconducting material. One of the most interesting features of this material is the occurrence of intrinsic Josephson effects in the c-axis transport properties.[27–29] The characteristic Josephson effects are observed in a single crystal of highly anisotropic High-T_c copper oxide superconductors, not in an artificial device made of them. It means that atomic-scale Josephson junction stacks are naturally realized there. The MQT in the High-T_c copper oxide compounds have been intensively studied from experimental[30–36] and theoretical[37–41] viewpoints.

2.2. Chiral superconductivity and topological ordered states

The next example is a spin-triplet chiral p-wave superconductor.[42,43] A fascinating feature is the presence of a non-trivial topological order. Also, hetero junctions composed of a kind of materials with "strong" spin-orbital coupling (e.g., topological insulators[44,45]) and conventional superconductors may provide a similar topological ordered state due to the superconducting proximity effect. Recently, various theories have predicted that a remarkable topological property can be realized in superconducting junction systems with topological insulators[46,47] or a kind of semiconductors (e.g., InAs).[48–50] These superconducting systems may give the physical realization of topological quantum computing.[51,52]

2.3. Multiband superconductivity

In 2008, a novel superconducting material was discovered.[9] This is an iron-pnictide superconductor, $LaFeAs(O_{1-x}F)$ whose superconducting transition temperature is 23 K. After the first discovery, the astonishing discovry rush of the iron-based compounds has come.[53-59] In contrast to High-T_c copper oxside superconductors, $3d$ electrons on the iron atom form the multiple bands in which the Cooper pairs condense into a multigap superconducting state. Various experiments revealed that multiple bands contribute to the superconductivity and multiple superconducting full (s-wave) gaps open below the transition temperature. See Refs. 60 and 61 for the experimental details. Another important example of the multiband superconducotor is magnesium diboride,[10] in which π- and σ-bands contribute to the superconductivity.

One of the peculiar features in multiband superconductivity is the presence of multiple superconducting order parameters (i.e., multigap superconductivity). It indicates that Josephson junctions with a multiband superconductor may have multiple tunneling channels between the two superconducting electrodes. Furthermore, a fluctuation between the multple superconducting gaps brings about a unique collective excitation, Leggett's mode.[21] The combination of these two facts suggest that Josephson effects been significantly altered.[11] Indeed, the classical Josephson junction features have been intensively studied. On the other hand, quantum application of the multiband superconductors and their Josephson junctions is an unexplored research field. The present work is a first contribution of Josephson junction with multiple tunneling channels to quantum engineering.

3. Model

Let us begin with the model for a hetero Josephson junction between single- and two-gap superconductors.[11,12,17-19] The system is formed by the two superconducting electrodes and an insulating layer with the length d and the area W sandwitched by them as shown in Fig. 1. Recently, similar Josephson junctions with iron-based superconductors[62-64] or magnesium diboride[65,66] have been successfully fabricated by several experimental groups. The DC current I_{ex} is applied to the junction as bias current. Since the top electrode has the two superconducting gaps, the present superconducting-insulator-superconducting (SIS) junction has the two tunneling channels. Therefore, the dynamics of the junction is described by two gauge invariant phase differences $\theta^{(1)}$ and $\theta^{(2)}$.[11,13] See fig. 2. The Josephson current density flowing into the interface is given by

$$J_{sc} = j_1 \sin \theta^{(1)} + j_2 \sin \theta^{(2)}, \tag{1}$$

in which j_i is the Josephson critical current density associated with the ith tunneling channel, while the voltage difference between the electrode v is related to the temporal evolution of $\theta^{(i)}$ through the Josephson relation [11]

$$\frac{\alpha_2}{\alpha_1 + \alpha_2}\dot{\theta}^{(1)} + \frac{\alpha_1}{\alpha_1 + \alpha_2}\dot{\theta}^{(2)} = \frac{2e\Lambda}{\hbar}v. \tag{2}$$

Magnitude of electric field coupling for the ith band $\alpha_i = \epsilon\mu_i/d$, where ϵ is the dielectric constant of the insulator and μ_i is the charge screening length for the ith band. The appearance of α_i and $\Lambda(= 1 + \alpha_1\alpha_2/(\alpha_1 + \alpha_2))$ in Eq. (2) are ascribed to the finite charge compressibility in the top superconducting electrode.[12] We assume that the corresponding region is distributed around the interface whose length is characterized by the charge screening length.

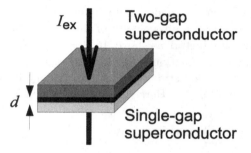

Fig. 1.   Schematic diagram for a SIS hetero Josephson junction.

4. Interband Josephson Coupling

We briefly review a key feature in multiband superconductivity, i.e., interband Josephson coupling. Such a kind of the interaction leads to a unique collective excitation mode in multiband superconductors. For two-band superconductors, there is a classical 1966 work by Leggett, who has shown that a two-band superconductor accommodates a special collective excitation, now called Leggett's mode, emerging as an out-of-phase mode between different superfluids.[21] The Leggett's mode plays an important role in the present formalism of the MQT in multiple-tunneling-channel Josephson junctions.

Let us start with a model for multiband superfluidity/superconductivity. We take the simplest possible BCS type Hamiltonian density for an N-band superfluid, assumed to be neutral here, is given as

$$\hat{\mathcal{H}}(\boldsymbol{r}) = \hat{\mathcal{H}}_{\text{kinetic}}(\boldsymbol{r}) + \hat{\mathcal{H}}_{\text{pairing}}(\boldsymbol{r}), \tag{3}$$

$$\hat{\mathcal{H}}_{\text{kinetic}} = \sum_{i=1}^{N} \sum_{\sigma=\uparrow,\downarrow} \hat{\psi}_{i\sigma}^{\dagger} \varepsilon_i(-i\nabla) \hat{\psi}_{i\sigma}, \tag{4}$$

$$\hat{\mathcal{H}}_{\text{pairing}} = - \sum_{i,j=1}^{N} g_{ij} \hat{\psi}_{i\uparrow}^{\dagger} \hat{\psi}_{i\downarrow}^{\dagger} \hat{\psi}_{j\downarrow} \hat{\psi}_{j\uparrow}, \tag{5}$$

where the dispersion $\varepsilon_i(\boldsymbol{k})$ is assumed to be parabolic with $\hbar = 1$. Here $\hat{\psi}_{i\sigma}$ ($i = 1, 2, \ldots, N$) is the field operator for the ith band in a superconductor, and $\boldsymbol{\mathcal{G}} \equiv (g_{ij})$ the pairing matrix, taken to be real and symmetric, and we drop its $\boldsymbol{k}$-dependence. While $\boldsymbol{\mathcal{G}}$ may come from (the short-range part of) the electron correlation, we ignore a long range Coulomb interaction since the mass of Leggett's modes is not affected by the Anderson-Higgs mechanism, although the group velocity of the mode reflects the Coulomb interaction in superconductors as shown in Ref. 22 for $N = 2$.

The grand partition function is given, with the imaginary-time functional integral method, in terms of a set of auxiliary fields $\Delta^{(i)}$ as $Z = Z_0 \int \prod_{i=1}^{N} \mathcal{D}\Delta^{(i)} \mathcal{D}\Delta^{(i)*} e^{-S_{\text{eff}}}$. Here the effective action is

$$S_{\text{eff}} = \int_0^{\beta} d\tau \int d^3\boldsymbol{r} \sum_{i,j=1}^{N} \Delta^{(i)*} (\boldsymbol{\mathcal{G}}^{-1})_{ij} \Delta^{(j)} - \sum_{i=1}^{N} \text{Tr} \ln \hat{G}_{0,i} - \sum_{i=1}^{N} \text{Tr} \ln \hat{G}_i^{-1}, \tag{6}$$

where we assume the inverse $\boldsymbol{\mathcal{G}}^{-1}$ is well-defined (i.e., $\boldsymbol{\mathcal{G}} > 0$), Z_0 the partition function for the non-interacting system, and β the inverse temperature. $\hat{G}_i$ is Green's function in the Nambu representation for the ith-band, which satisfies

$$\hat{G}_i^{-1} = \hat{G}_{0,i}^{-1}(\hat{I} - \hat{G}_{0,i}\hat{K}_i), \tag{7}$$

where $\hat{G}_{0,i}$ is the free-fermion Green's function, $\hat{I}$ the unit matrix, and $\hat{K}_i = \sigma_1 \text{Re}\Delta^{(i)} - i\sigma_2 \text{Im}\Delta^{(i)}$, with σ_α being the Pauli matrices. The gaps are coupled through the inverse of the pairing matrix $(\boldsymbol{\mathcal{G}}^{-1})$, and we shall see that the non-zero off-diagonal elements of $\boldsymbol{\mathcal{G}}^{-1}$ determine the Leggett's modes.

With this effective action the static gap equation reads

$$\Delta_i = \sum_{j=1}^{N} g_{ij} \Delta_j N_j \int_0^{\omega_c} d\xi \frac{\tanh(\beta E_j/2)}{E_j}, \tag{8}$$

where $E_i \equiv \sqrt{\xi^2 + |\Delta_i|^2}$, ω_c a cut-off frequency, and N_j the density of states (DOS) of the jth fermion on the Fermi surface. Here we assume that each of the $\Delta^{(i)}$'s is constant (an s-wave). We can then look at the phase $\varphi_0^{(i)}$ in $\Delta_i = |\Delta_i| e^{i\varphi_0^{(i)}}$. The gap functions on different bands can take either the same or opposite signs, i.e., the phase difference, $\phi_0^{(i,j)} \equiv \varphi_0^{(j)} - \varphi_0^{(i)}$, takes either 0 or $\pm\pi$ and should obviously satisfy

$$\sum_{i=1}^{N} \phi_0^{(i,i+1)} \equiv 0 \quad (\text{mod } 2\pi). \tag{9}$$

with $\varphi_0^{(N+1)} \equiv \varphi_0^{(1)}$.

Let us derive an effective action for the fluctuations in the superfluid phase to single out the collective dynamics at zero temperature. We first decompose the phase $\varphi_0^{(i)} + \varphi^{(i)}$, into the equilibrium $\varphi_0^{(i)}$ (as obtained in the mean-field gap equation) and the phase fluctuation $\varphi^{(i)}$. Around the solution of Eq. (8), we obtain an action,

$$S_{\text{eff}} = \int_0^\beta d\tau \int dr V_{\text{interband}} + \sum_{i=1}^{N} \sum_{m=1}^{\infty} \frac{1}{m} \text{Tr}(\hat{G}_{0,i}' \hat{K}_i')^m, \tag{10}$$

where

$$V_{\text{interband}} = \frac{1}{2} \sum_{i<j} \eta_{ij} \lambda_{ij} [1 - \cos(\varphi^{(j)} - \varphi^{(i)})], \tag{11}$$

$$\lambda_{ij} = 4|(\boldsymbol{G}^{-1})_{ij}||\Delta_i||\Delta_j|, \tag{12}$$

$$\eta_{ij} = \cos(\phi_0^{(i,j)} + \kappa_{ij}), \tag{13}$$

comprises a sum of Josephson-couplings between the phases of different superfluid components that represents the interband Josephson currents caused by the relative phase fluctuations. In the above expression, the constraint $\phi_0^{(i,j)} = 0$ or $\pm\pi$ is used, and $\kappa_{ij} (= 0$ or $\pi)$ the sign of $(\boldsymbol{G}^{-1})_{ij}$, i.e., $e^{i\kappa_{ij}} \equiv -\text{sgn}(\boldsymbol{G}^{-1})_{ij}$. The primed quantities are gauge-transformed

$$\hat{G}_{0,i}'(x; x') = \hat{U}_i(x)\hat{G}_{0,i}(x; x')\hat{U}_i^\dagger(x'), \tag{14}$$

where $\hat{U}_i = e^{-i\sigma_3(\varphi_0^{(i)} + \varphi^{(i)})/2}$ and $x \equiv (\tau, \boldsymbol{r})$, with the Dyson equation transformed into

$$\hat{G}_i'^{-1} = \hat{G}_{0,i}'^{-1}(\hat{I} - \hat{G}_{0,i}'\hat{K}_i'). \tag{15}$$

When $N = 2$, Eq. (11) reduces to $V_{\text{interband}} = (\eta_{12}\lambda_{12}/2)[1 - \cos(\varphi^{(2)} - \varphi^{(1)})]$ with a single inter-band Josephson coupling. The sign of $-(\boldsymbol{G}^{-1})_{12}(= g_{12}/\det \boldsymbol{G})$ is equal to that of g_{12}. Hence $\kappa_{12} = 0$ when $g_{12} > 0$, for which

$\phi_0^{(1,2)} = 0$ for $V_{\text{interband}}$ to give a stable gap solution according to Eq. (13). Similarly, $\kappa_{12} = \phi_0^{(1,2)} = \pi$ when $g_{12} < 0$. Thus we always have $\eta_{12} = 1$, which implies that the sign of g_{12} is totally irrelevant to the spectrum of collective modes in the two-band case. In other words the sign of g_{12} and the value of $\phi_0^{(1,2)}$ are a matter of convention for Leggett's mode, as noted in Refs. 21 and 22. We remark that the situation becomes completely different if $N \geq 3$. In this case, a new quantity to classify the Leggett's mode is necessary. This point is closely discussed in Ref. 23.

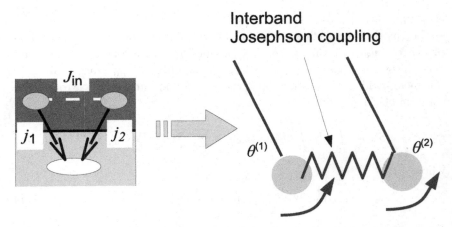

Fig. 2. Schematic diagram for a Josephson junction with multiple tunneling channels. The two conventional Josephson currents (j_1 and j_2) flow while a novel coupling, interband Josephson coupling (J_{in}) leads to interaction between the two tunneling channels. As a result, the system is well described by coupled non-linear oscillators.

5. Classical Phase Dynamics

We show the (classical) Lagrangian for $\theta^{(1)}$ and $\theta^{(2)}$. On the basis of Refs. 11 and 13, we find that the Lagrangian with the bias current (and without the external magnetic field) in the real time formalism is given by

$$L = \frac{1}{2} \frac{\hbar^2 C}{(2e)^2} \left(\frac{\dot{\theta}^2}{\Lambda} + \frac{\dot{\psi}^2}{\alpha_1 + \alpha_2} \right) - V + E_c \frac{I_{\text{ex}}}{I_c} \theta, \tag{16a}$$

$$V = -E_{J1} \cos \theta^{(1)} - E_{J2} \cos \theta^{(2)} - \kappa E_{\text{in}} \cos \psi, \tag{16b}$$

where

$$\theta = \frac{\alpha_2}{\alpha_1 + \alpha_2} \theta^{(1)} + \frac{\alpha_1}{\alpha_1 + \alpha_2} \theta^{(2)}, \quad \psi = \theta^{(1)} - \theta^{(2)}.$$

θ is regarded as the center-of-mass variable, while ψ the relative one. The temporal derivative of θ is equivalent to the left hand side of Eq. (2). The capacitance and the Josephson energy for the ith tunneling channel are, respectively, $C = \epsilon W/4\pi d$ and $E_{Ji} = \hbar W j_i/2e$. The third term in Eq. (16b), which is characteristic coupling in multiband superconductors, is the interband Josephson energy and $E_{in} = \hbar W |J_{in}|/2e$. J_{in} is determined by the superconducting gap amplitudes and the magnitude of superconducting pair potential.[23] The factor κ indicates the sign of J_{in}, $\kappa = J_{in}/|J_{in}|$. The last term in Eq. (16a) is the contribution from the bias current, in which $I_c = W|j_1 + \kappa j_2|$ and $E_c = \hbar I_c/2e$.

Let us derive the classical equation of motion from Eq. (16a) to obtain intuitive understanding for the present system. The Euler-Lagrange equation for the center-of-mass variable θ is

$$\frac{\hbar^2 C}{(2e)^2} \Lambda^{-1} \ddot{\theta} - \left(-E_{J1} \sin\theta^{(1)} - E_{J2} \sin\theta^{(2)} + E_c \frac{I_{ex}}{I_c} \right) = 0.$$

Then

$$\Lambda^{-1} \ddot{\theta} + \omega_{P1}^2 \sin\theta^{(1)} + \omega_{P2}^2 \sin\theta^{(2)} = \omega_c^2 \frac{I_{ex}}{I_c}, \tag{17}$$

where $\hbar\omega_{Pi} = \sqrt{2E_C E_{Ji}}$ and $\hbar\omega_c = \sqrt{2E_C E_c}$. Here, we emply the conventional expression for the charging energy $E_C = (2e)^2/2C$. One finds that the above equation is regarded as the Maxwell equation. In addition, ω_{Pi} corresponds to the Josephson-plasma frequency for the ith tunneling channel. On the other hand, the Euler-Lagrange equation for ψ is

$$0 = \frac{\hbar^2 C}{(2e)^2} \frac{1}{\alpha_1 + \alpha_2} \ddot{\psi} - \left(-\frac{\alpha_1}{\alpha_1 + \alpha_2} E_{J1} \sin\theta^{(1)} \right.$$
$$\left. + \frac{\alpha_2}{\alpha_1 + \alpha_2} E_{J1} \sin\theta^{(2)} - \kappa E_{in} \sin\psi \right).$$

Then we have

$$\ddot{\psi} + \kappa\omega_{JL}^2 \sin\psi = -\alpha_1 \omega_{J1}^2 \sin\theta^{(1)} + \alpha_2 \omega_{J2}^2 \sin\theta^{(2)}. \tag{18}$$

The quantity ω_{JL} is the frequency of the Josephosn-Leggett mode,[11,13,21,22] defined by $\hbar\omega_{JL} = \sqrt{2(\alpha_1 + \alpha_2)E_C E_{in}}$. These coupled equations indicate that the external bias current directly induces the time evolution of θ but not for ψ. The Josephson relation (2) also supports this argument. The dynamical behavior of ψ is excited by coupling with the center-of-mass variable although ψ is not directly interacted with electric field. We also mention that the present system has two distinct energy scales, the Josephson-plasma ω_{Ji} and the Josephson-Leggett ω_{JL}.

6. MQT Escape Rate

We now turn to a quantum aspect of the Josephson junction with multiple tunneling channel, i.e., to evaluate the MQT escape rate. Hereafter, we concentrate on the case of $\kappa > 0$. A similar calculation is possible for $\kappa < 0$.

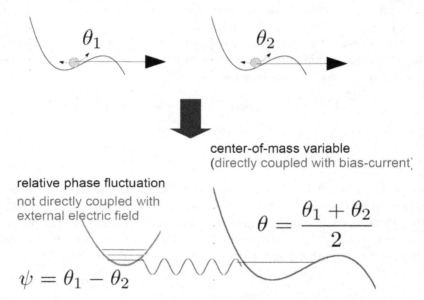

Fig. 3. Schematics for a tunneling process taken into account in this paper. The top panel shows a simultaneous tunneling process with respect to $\theta^{(1)}$ and $\theta^{(2)}$. This process is interpreted as a process depicted in the bottom panel.

In the present paper, we concentrate on a spesific transition process in which both of the phase variables $\theta^{(1)}$ and $\theta^{(2)}$ simultaneously escape from a potential well, as shown in Fig. 3. According to Ref. 4, the expression for the MQT escape rate with respect to a transition process $|\theta = 0, \psi = 0\rangle \to |0, 0\rangle$ is given by

$$\Gamma = \frac{2}{\hbar\beta}\text{Im}\,K(\{0\}, \{0\}; \beta). \tag{19}$$

The symbol $\{0\}$ means that $(\theta, \psi) = (0, 0)$. β is the inverse temperature. The propagator in the imaginary time path-integral formula $K(X, X'; \beta)$ is given as

$$K(X, X'; \beta) = \int_{X(0)=X'}^{X(\hbar\beta)=X} \mathcal{D}\theta \mathcal{D}\psi \, e^{-\int_0^{\hbar\beta} d\tau \, L^E/\hbar},$$

where $X = (\theta, \psi)$ and $\mathcal{L}_E$ is the Euclidean Lagrangian for Eq. (16a). The value of ψ should be rigidly pinned due to the large inter-band Josephson coupling. When $\kappa > 0$, this value should be 0. Using small ψ-expantion, we obtain an approximate formula for the Euclidean Lagrangian up to an irrelevant constant term $L^E = L_{cm}^E + L_{rlt}^E + L_{int}^E$, where

$$L_{cm}^E = \frac{\hbar^2}{4E_C} \left(\frac{d\theta}{d\tau}\right)^2 - E_J \left(\cos\theta + \frac{I_{ex}}{I_c}\theta\right), \tag{20a}$$

$$L_{rlt}^E = \frac{\hbar^2}{4(\alpha_1 + \alpha_2)E_C} \left(\frac{d\psi}{d\tau}\right)^2 + \frac{1}{2}E_{in}\psi^2, \tag{20b}$$

$$L_{int}^E = g_+ E_J \psi^2 \cos\theta - g_- E_J \psi \sin\theta. \tag{20c}$$

Here, we emply $E_J \equiv E_c = E_{J1} + E_{J2}$ for $\kappa > 0$ and $\Lambda \approx 1$. The coupling constants g_+ and g_- are, respectively, given by $g_+ = (E_{J1}/2E_J)[\alpha_1/(\alpha_1 + \alpha_2)]^2 + (E_{J2}/2E_J)[\alpha_2/(\alpha_1 + \alpha_2)]^2$ and $g_- = (E_{J1}/E_J)[\alpha_1/(\alpha_1 + \alpha_2)] - (E_{J2}/E_J)[\alpha_2/(\alpha_1 + \alpha_2)]$. We emphasize that a quantum fluctuation of ψ is allowed even though the classical variable ψ is rigidly fixed. To take into account of a quantum fluctuation of ψ to Γ, we employ the mean field approximation for ψ. We replace ψ^2 and ψ with the expectation values with respect to the thermal equilibrium state described by L_{rlt}^E. We find that $\langle\psi\rangle_{th} = 0$, while at zero temperature limit

$$\langle\psi^2\rangle_{th} = \frac{\hbar}{2m_{rlt}\omega_{JL}}, \quad m_{rlt} = \frac{\hbar^2}{2(\alpha_1 + \alpha_2)E_C},$$

which corresponds to the zero-point fluctuation of the "quantized" Josephson-Leggett mode. The detail evaluation of $\langle\psi\rangle_{th}$ and $\langle\psi^2\rangle_{th}$ is shown in Appendix. Thus, the system is described in term of θ, and the effective Lagrangian is given by

$$L_{cm,eff}^E = \frac{\hbar^2}{4E_C} \left(\frac{d\theta}{d\tau}\right)^2 + V_{cm,eff}, \tag{21a}$$

where $V_{cm,eff}$ is the renormalized potential

$$V_{cm,eff} = -E_J \left[(1-\varepsilon)\cos\theta + \frac{I_{ex}}{I_c}\theta\right], \tag{21b}$$

$$\varepsilon = g_+ \langle\psi^2\rangle_{th} \approx \frac{g_+}{\sqrt{2}}(\alpha_1 + \alpha_2)\frac{\omega_P}{\omega_{JL}}\sqrt{\frac{E_C}{E_J}}. \tag{21c}$$

The (net) Josephson-plasma frequency of the present junction for $\kappa > 0$ is $\omega_P = \sqrt{2E_C E_J}/\hbar$. Then, we have $K(\{0\}, \{0\}; \beta) \approx \int_{\theta(0)=0}^{\theta(\hbar\beta)=0} \mathcal{D}\theta \, \exp(-\hbar^{-1} \int_0^{\hbar\beta} L^{\mathrm{E}}_{\mathrm{cm,eff}} d\tau)$, and we can evaluate Eq. (19) using the standard instanton method.[2-4] Hence, the MQT escape rate with the zero-point fluctuation of ψ is given by

$$\Gamma = 12\omega_P(I)\sqrt{\frac{3V_0}{2\pi\hbar\omega_P(I)}} \exp\left(-\frac{36V_0}{5\hbar\omega_P(I)}\right), \qquad (22)$$

where $\omega_P(I) = \omega_P[(1 - \varepsilon)^2 - I^2]^{1/4}$, $V_0 = \hbar^2[\omega_P(I)]^2 \cot^2\theta_0/3E_C$, $(1 - \varepsilon)\sin\theta_0 = I$, and $I = I_{\mathrm{ex}}/I_{\mathrm{c}}$. The ration of the above formula to the MQT rate without the correction is

$$\frac{\Gamma}{\Gamma_0} = \frac{\omega_P(I)}{\omega_P(I; \varepsilon = 0)}\sqrt{\frac{V_0}{\hbar\omega_P(I)}\frac{\hbar\omega_P(I; \varepsilon = 0)}{V_0(\varepsilon = 0)}}$$

$$\times \exp\left[-\frac{36}{5}\left(\frac{V_0}{\hbar\omega_P(I)} - \frac{V_0(\varepsilon = 0)}{\hbar\omega_P(I; \varepsilon = 0)}\right)\right]$$

where Γ_0 can be obtained taking the limit $\varepsilon \to 0$ in Γ. We remark that when $1 > \varepsilon > 0$

$$\omega_P(I) < \omega_P(I; \varepsilon = 0), \qquad \frac{V_0}{\hbar\omega_P(I)} < \frac{V_0(\varepsilon = 0)}{\hbar\omega_P(I; \varepsilon = 0)}.$$

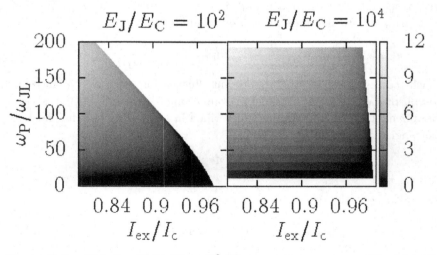

Fig. 4.   Ratio of Γ to Γ_0 with $E_J/E_C = 10^2$. We assume that $\alpha_1 = \alpha_2 = 0.1$ and $j_1 = j_2$ for simplicity.

Figure 4 shows the ratio of Γ to Γ_0, which is the MQT escape rate without any correction ψ (i.e., $\varepsilon = 0$), for various values of the parameters I_{ex}/I_c and ω_P/ω_{JL}. We find that drastic enhancement occurs in vast area on the parameter space. In particular, such enhancement is pronounced at the region for large ω_P/ω_{JL} (i.e., low Josephson-Leggett mode's frequency region). The mechanism is explained from the observation of Eq. (21b). The contribution from the zero-point fluctuation with respect to ψ changes the Josephson coupling energy (or, the Josephson critical current). Here, we find that $1 - \varepsilon < 1$ since Eq. (21c) implies $\varepsilon > 0$. Accordingly, the zero-point fluctuation always decreases the Josephson coupling energy. It means that the barrier height of the renormalized potential is effectively lower than that of the bare potential. In fact, we find that for fixed I the factor $R(\varepsilon) \equiv V_0/\hbar\omega_P(I)$ in the exponent of Eq. (22), which is the ratio of the potential barrier height to "incident" energy in the corresponding tunneling problem, is smaller than $R(\varepsilon = 0)$ when $0 < \varepsilon < 1$. In addition, the zero-point fluctuation becomes large as the frequency reduces. Thus, the considerable enhancement occurs for case of low ω_{JL}. In the conventional system, MQT is completely determined by the magnitude of E_J/E_C, which is an important parameter for designing a superconducting Josephson qubit.[6,8] The present system, on the ohter hand, is characterized by not only such a quantity but also $\omega_P/\omega_{JL}(\propto E_J/E_{in})$.

The present paper focuses on a specific transition process $|\theta\rangle = 0, \psi = 0 \rightarrow |0, 0\rangle$ to clarify the effect of the Josephson-Leggett mode, as shown in Eq. (19). We note that this process is not a unique candidate for evaluating the MQT since this system contains the two degrees of freedom $\theta^{(1)}$ and $\theta^{(2)}$. A possible situation is that one of the tunneling channel is completely a running state while the phase variable corresponding to the other is still confined in a potential as a metastable state.[67] We may express the corresponding transition process as $|\theta^{(1)} = 0\rangle \rightarrow |0\rangle$ with $\theta^{(2)} = f(t)$ and $j_1 \gg j_2$, where $f(t)$ is a time-dependent c-number function. This case is similar to MQT with a periodic time-dependent function in Ref.[68] Alternatively, ψ may be regarded as a non-measured quantity, i.e., environment since it is not directly coupled with an external exlecric field. It is possible to evaluate the MQT rate tracing over the relative variable ψ, like Refs.[1,69,70] Then, we encounter an interesting situation that competition between the enhancement by the zero-point fluctuation and "dissipation" occurs. We also mention that extension of the present theory to intrinsic Josephson junction arrays with multiple gaps is an intriguing issue since some layered iron-based superconductors have very large anisotropy like High-T_c copper oxide superconductors [71] and an experiment [72] suggests coherent transport in the direction of c-axis. It is an interesting work to reveal a cooperative phenomenon between the Josephson-Leggett mode and characteristic coupling in the intrinsic Josephson junction states, i.e., capacitive coupling [39] and inductive coupling.[41]

7. Summary

In summary, we have constructed a theory of the MQT in hetero Josephson junctions with multiple tunneling channels. We have clarified that the zero-point fluctuation of the relative phase differences brings about the drastic enhancement of the MQT escape rate. The enhancement is larger when the Leggett's mode has a lighter mass than that of the Josephson-plasma.

Acknowledgments

YO thanks to illuminating discussion with M. Nakahara, S. Kawabata, and Y. Chizaki. TK was partially supported by Grant-in-Aid for Scientific Research (C) (No. 22540358) from Japan Science and Technology Agency.

Appendix A. Mean Field Approximation for Out-of-Phase Variable

In This Appendix, We Follow The Evaluation of $\langle \psi \rangle_{\text{th}}$ and $\langle \psi^2 \rangle_{\text{th}}$. First of all, we define the expectation value $\langle A[\psi] \rangle_{\text{th}}$ for any functional A with respect to ψ with the functional integral formula as

$$\langle A[\psi] \rangle_{\text{th}} = \frac{1}{Z_{\psi,0}} \int_{\psi(0)=0}^{\psi(\hbar\beta)=0} [\mathrm{d}\psi] A[\psi] \exp\left(-\frac{1}{\hbar} \int_0^{\hbar\beta} L_{\text{rlt}}^{\text{E}} \right). \qquad (A.1)$$

where

$$Z_\psi[J] = \int_{\psi(0)=0}^{\psi(\hbar\beta)=0} [\mathrm{d}\psi] \exp\left(-\frac{1}{\hbar} \int_0^{\hbar\beta} L_{\text{rlt},J}^{\text{E}} \right), \qquad (A.2)$$

$$Z_{\psi,0} = Z_\psi[J=0], \qquad (A.3)$$

$$L_{\text{rlt},J}^{\text{E}} = L_{\text{rlt}}^{\text{E}} + J(\tau)\psi. \qquad (A.4)$$

We find that

$$Z_\psi[J] = \frac{1}{2\sinh(\omega_{\text{JL}}\hbar\beta/2)} \exp\left(-\frac{1}{\hbar} S_{\text{rlt},c}^{\text{E}}[J] \right), \qquad (A.5)$$

where

$$S_{\text{rlt,c}}^{\text{E}}[J] = -\frac{m_{\text{rlt}}}{2} \int_0^{\hbar\beta} \mathrm{d}\tau \int_0^{\hbar\beta} \mathrm{d}\tau' \frac{-J(\tau)}{m_{\text{rlt}}} G_\psi(\tau - \tau') \frac{-J(\tau')}{m_{\text{rlt}}}$$

$$G_\psi(\tau) = \frac{1}{2\omega_{\rm JL}\sinh(\omega_{\rm JL}\hbar\beta/2)}\left[\theta(\tau)\cosh\omega_{\rm JL}\left(\frac{\hbar\beta}{2}-\tau\right)\right.$$
$$\left.+\theta(-\tau)\cosh\omega_{\rm JL}\left(\frac{\hbar\beta}{2}+\tau\right)\right]$$

Therefore, we have

$$\langle\psi(\tau)\rangle_{\rm th} = 0,$$

$$\langle T_\tau\psi(\tau)\psi(\tau')\rangle_{\rm th} = \frac{\hbar}{2m_{\rm rlt}}\frac{1}{2\omega_{\rm JL}\sinh(\omega_{\rm JL}\hbar\beta/2)}[G_\psi(\tau-\tau')+G_\psi(\tau'-\tau)],$$

$$\langle\psi^2(\tau)\rangle_{\rm th} = \frac{\hbar}{2m_{\rm rlt}\omega_{\rm JL}}\cot\frac{\omega_{\rm JL}\hbar\beta}{2} \to \frac{\hbar}{2m_{\rm rlt}\omega_{\rm JL}} \quad (\beta\to\infty).$$

References

1. A. O. Caldeira and A. J. Leggett, *Phys. Rev. Lett.* **46**, 211 (1981).
2. S. Coleman, *Aspects of Symmetry* (Cambridge University Press, Cambridge, UK, 1985) Chap.7.
3. R. Rajaraman, *Solitons and Instantons: An Introduction to Solitons and Instantons in Quantum Field Theory* (Elsevier, Amsterdam, 1987).
4. E. Šimánek, *Inhomogeneous Superconductors: Granular and Quantum Effects* (Oxford University Press, New York, 1994) Chap.4.
5. A. Larkin and A. Varkamov, *Theory of Fluctuations in Superconductors* Revised Edition (Oxford University Press, New York, 2005) Chap.13.
6. J. Q. You and F. Nori, *Physics Today* **58**, 42 (2005).
7. J. Clarke and F. Wilhelm, *Nature* **453**, 1031 (2008).
8. M. Nakahara and T. Ohmi, *Quantum Computing: From Linear Algebra To Physical Realizations* (CRC Press, Talyor & Francis Group, Boca Raton, 2008) Chap.15.
9. Y. Kamihara, T. Watanabe, M. Hirano and H. Hosono, *J. Am. Chem. Soc.* **130**, 3296 (2008).
10. X. X. Xi, *Rep. Prog. Phys.* **71**, 116501 (2008).
11. Y. Ota, M. Machida, T. Koyama and H. Matsumoto, *Phys. Rev. Lett.* **102**, 237003 (2009).
12. Y. Ota, M. Machida and T. Koyama, *Phys. Rev. B* **82**, 140509(R) (2010).
13. T. Koyama, Y. Ota and M. Machida, *Physica C* **470** (2010) 1481.
14. Y. Ota, N. Nakai, H. Nakamura, M. Machida, D. Inotani, Y.Ohashi, T. Koyama and H. Matsumoto, *Phys. Rev. B* **81**, 214511 (2010).
15. Y. Ota, M. Machida, T. Koyama and H. Matsumoto, *Phys. Rev. B* **81**, 014502 (2010).
16. Y. Ota, M. Machida and T. Koyama, *J. Phys. Soc. Jpn.* **78**, 103701 (2009).
17. D. F. Agterberg, E. Demler and B. Janko, *Phys. Rev. B* **66**, 214507 (2002).
18. D. Inotani and Y. Ohashi, *Phys. Rev. B* **79**, 224527 (2009).
19. J. Linder, I. B. Sperstad and A. Sudbø, *Phys. Rev. B* **80**, 020503(R) (2009).
20. Y. Ota, M. Machida and T. Koyama, arXiv:1010.2804; Accepted for publication in *Phys. Rev. B*.

208

21. A. J. Leggett, *Prog. Theor. Phys.* **36**, 901 (1966).

22. S. G. Sharapov, V. P. Gusynin and H. Beck, *Eur. Phys. J. B* **30**, 45 (2002).

23. Y. Ota, M. Machida, T. Koyama and H. Aoki, arXiv:1008.3212; Accepted for publication in *Phys. Rev. B*.

24. M. Tinkham *Introduction to Superconductivity* 2nd ed. (Dover, New York, 2004).

25. M. A. Sillanpää, J. I. Park and R. Simmonds, *Science* **449**, 438 (2007).

26. K. Machida, M. Ichioka, M. Takigawa and N. Nakai, *Unconventional Superconductivity with Either Multi-component or Multi-band, or with Chirality* in *Ruhenate and Rutheno-Cuprate Materilas*, edited by C. Noce, A. Vecchione, M. Cucco and A. Romano (Springer, Berlin, 2002) pp.242-pp.254.

27. R. Kleiner, F. Steinmeyer, G. Kunkel and P. Müller, *Phys. Rev. Lett.* **68**, 2394 (1992).

28. G. Oya, N. Aoyama, A. Irie, S. Kishida and H. Tokutaka, *Jpn. J. Appl. Phys.* **31** L829 (1992).

29. K. Tamasaku, Y. Nakamura and S. Uchida, *Phys. Rev. Lett.* **69**, 1455 (1992).

30. T. Bauch, F. Lombardi, F. Tafuri, A. Barone, G. Rotoli, P. Delsing and T. Claeson, *Phys. Rev. Lett.* **94**, 087003 (2005).

31. K. Inomata, S. Sato, K. Nakajima, A. Tanaka, Y. Takano, H. B. Wang, M. Nagao, H. Hatano and S. Kawabata, *Phys. Rev. Lett.* **95**, 107005 (2005).

32. X. Y. Jin, J. Lisenfeld, Y. Koval, A. Lukashenko, A. V. Ustinov and P. Müller, *Phys. Rev. Lett.* **96**, 177003 (2006).

33. H. Kashiwaya, T. Matsumoto, H. Shibata, S. Kashiwaya, H. Eisaki, Y. Yoshida, S. Kawabata and Y. Tanaka, *J. Phys. Soc. Jpn.* **77**, 104708 (2008).

34. K. Ota, K. Hamada, R. Takemura, M. Ohmaki, T. Machi, K. Tanabe, M. Suzuki, A. Maeda and H. Kitano, *Phys. Rev. B* **79**, 134505 (2009).

35. I. Kakeya, K. Hamada, T. Tachiki, T. Watanabe, and M Suzuki, *Supercond. Sci. Technol.* **22** 114014 (2009).

36. H. Kashiwaya, T. Matsumoto, H. Shibata, H. Eisaki, Y. Yoshida, H. Kambara, S. Kawabata and S. Kashiwaya, *Appl. Phys. Express* **3**, 043101 (2010).

37. T. Koyama, *J. Phys. Soc. Jpn.* **70**, 2114 (2001).

38. S. Kawabata, S. Kashiwaya, Y. Asano and Y. Tanaka *Phys. Rev. B* **70**, 132505 (2004).

39. M. Machida and T. Koyama, *Supercond. Sci. Technol.* **20**, S23 (2007).

40. S. Kawabata, A. A. Golubov, Ariando, C. J. M. Verwijs, H. Hilgenkamp and J. R. Kirtley, *Phys. Rev. B* **76**, 064505 (2007).

41. S. Savel'ev, A. O. Sboychakov, A. L. Rakhmanov and F. Nori, *Phys. Rev. B* **77**, 014509 (2008).

42. T. M. Rice and M. Sigrist, *J. Phys.: Cond. Matter* **7**, L643 (1995).

43. A. P. Mackenzie and Y. Maeno, *Rev. Mod. Phys.* **75**, 657 (2003).

44. M. Z. Hasan and C. L. Kane, *Rev. Mod. Phys.* **82**, 3045 (2010).

45. X.-L. Qi and S.–C. Zhang, arXiv:1008.2026 (unpublished).

46. L. Fu and C. L. Kane, *Phys. Rev. Lett.* **100**, 096407 (2008).

47. L. Fu and C. L. Kane, *Phys. Rev. B* **79**, 161408(R) (2009).

48. R. M. Lutchyn, J. D. Sau and S. Das Sarma, *Phys. Rev. Lett.* **105**, 07701 (2010).

49. J. Linder, Y. Tanaka, T. Yokoyama, A. Sudbø and N.Nagaosa, *Phys. Rev. Lett.* **104**, 067001 (2010).

50. Y. Oreg, G. Refael and F. von Oppen, *Phys. Rev. Lett.* **105**, 177002 (2010).

51. C. Nayak, S. H. Simon, A. Stern, M. Freedman and S. Das Sarma , *Rev. Mod. Phys.* **80**, 1083 (2008).

52. P. Bonderson, S. Das Sarma, M. Freedman and C. Nayak, arXiv:1003.2856 (unpublished).

53. M. Rotter, M. Tegel and D. Johrendt, *Phys. Rev. Lett.* **101**, 107006 (2008).

54. K. Sasmal, B. Lv, B. Lorenz, A. M. Guloy, F. Chen, Y.-Y. Xue and C.-W. Chu, *Phys. Rev. Lett* **101**, 107007 (2008).

55. R. Zhi-An, L. Wei, Y. Jie, Yi. Wei, S. Xiao-Li, L. Zheng-Cai, C. Guang-Can, D. Xiao-Li, S. Li-Ling, Z. Fang and Z. Zhong-Xian, *Chin. Phys. Lett.* **25**, 2215 (2008).

56. J. H. Tapp, Z. Tang, B. Lv, K. Sasmal, B. Lorenz, P. C. W. Chu and A. M. Guloy, *Phys. Rev. B* **78**, 060505(R) (2008).

57. X. C. Wang, Q. Liu, Y. Lv, W. Gao, L. X. Yang, R. C. Yu, F. Y. Li and C. Jin, *Solid. State. Commun.* **148**, 538 (2008).

58. F.-C. Hsu, J.-Y. Luo, K.-W. Yeh, T.-K. Chen, T.-W. Huang, P. M. Wu, Y.-C. Lee, Y.-L. Huang, Y.-Y. Chu, D.-C. Yan and M.-K. Wu, *Proc. Natl. Acad. Sci. U.S.A.* **105**, 14262 (2008).

59. H. Ogino, Y. Matsumura, Y. Katsura, K. Ushiyama, S. Horii, K. Kishio and J. Shimoyama, *Supercond. Sci. Technol.* **22**, 075008 (2009).

60. K. Ishida, Y. Nakai and H. Hosono, *J. Phys. Soc. Jpn.* **78**, 062001 (2009).

61. J. Paglione and R. L. Greene, *Nature Phys.* **6**, 645 (2010).

62. X. Zhang, Y. S. Oh, Y. Liu, L. Yan, K. H. Kim, R. L. Greene and I. Takeuchi, *Phys. Rev. Lett.* **102**, 147002 (2009).

63. C. T. Wu, H. H. Chang, J. Y. Luo, T. J. Chen, F. C. Hsu, T. K. Chen, M. J. Wang and M. K. Wu, *Appl. Phys. Lett.* **96**, 122506 (2010).

64. S. Schmidt, S. Döring, F. Schmidl, V. Grosse, P. Seidel, K. Iida, F. Kurth, S. Haindl, I. Mönch and B. Holzapfel, *Appl. Phys. Lett.* **97**, 172504 (2010).

65. Z.-Z. Li, Y. Xuan, H.-J. Tao, Z.-A. Ren, G.-C. Che, B.-R. Zhao and Z.-X. Zhao, *Supercond. Sci. Technol.* **14**, 944 (2001).

66. H. Shunakage, K. Tsujimoto, Z. Wang and M. Tonouchi, *Supercond. Sci. Technol.* **17**, 1376 (2004).

67. Y. Chizaki and S. Kawabata (private communication).

68. M. P. A. Fisher, *Phys. Rev. B* **37**, 75 (1988).

69. M. V. Fistul, *Phys. Rev. B* **75**, 014502 (2007).

70. S. Kawabata, T. Bauch and T. Kato, *Phys. Rev. B* **80** 174513 (2009).

71. H. Nakamura, M. Machida, T. Koyama and N. Hamada, *J. Phys. Soc. Jpn.* **78**, 123712 (2008).

72. H. Kashiwaya, K. Shirai, T. Matsumoto, H. Shibata, H. Kambara, M. Ishikado, H. Eisaki, A. Iyo, S. Shamoto, I. Kurosawa and S. Kashiwaya, *Appl. Phys. Lett.* **96**, 202504 (2010).